REVISE EDEXCEL GCSE (9–1)
Chemistry
Higher

REVISION GUIDE

Series Consultant: Harry Smith

Author: Nigel Saunders

D0318382

A note from the publisher

In order to ensure that this resource offers high-quality support for the associated Pearson qualification, it has been through a review process by the awarding body. This process confirms that this resource fully covers the teaching and learning content of the specification or part of a specification at which it is aimed. It also confirms that it demonstrates an appropriate balance between the development of subject skills, knowledge and understanding, in addition to preparation for assessment.

Endorsement does not cover any guidance on assessment activities or processes (e.g. practice questions or advice on how to answer assessment questions), included in the resource nor does it prescribe any particular approach to the teaching or delivery of a related course.

While the publishers have made every attempt to ensure that advice on the qualification and its assessment is accurate, the official specification and associated assessment guidance materials are the only authoritative source of information and should always be referred to for definitive guidance.

Pearson examiners have not contributed to any sections in this resource relevant to examination papers for which they have responsibility.

Examiners will not use endorsed resources as a source of material for any assessment set by Pearson.

Endorsement of a resource does not mean that the resource is required to achieve this Pearson qualification, nor does it mean that it is the only suitable material available to support the qualification, and any resource lists produced by the awarding body shall include this and other appropriate resources.

Question difficulty
Look at this scale next to each exam-style question. It tells you how difficult the question is.

For the full range of Pearson revision titles across KS2, KS3, GCSE, Functional Skills, AS/A Level and BTEC visit:
www.pearsonschools.co.uk/revise

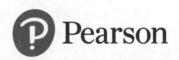
Pearson

Contents

- - - - - - - - - - - - - - - - - -

A small bit of small print:
Edexcel publishes Sample Assessment Material and the Specification on its website. This is the official content and this book should be used in conjunction with it. The questions in Now try this have been written to help you practise every topic in the book. Remember: the real exam questions may not look like this.

Formulae

You should be able to recall the formulae of elements, simple compounds and ions.

Elements

An **element** is a substance made from atoms with the same number of protons in the nucleus. Each element has its own **chemical symbol**, which:

- consists of one, two or three letters
- starts with a capital letter
- has any other letters in lower case.

For example, mercury is Hg. You can find the symbols for the elements in the periodic table.

Atoms and molecules

An **atom** is the smallest particle of an element that still has its chemical properties.

A **molecule** consists of two or more atoms chemically joined together.

Most elements that are gases at room temperature exist as molecules with two atoms. They are shown by **chemical formulae**, e.g. H_2, N_2, O_2, F_2, Cl_2.

Compounds

A **compound** consists of two or more different elements chemically joined together.

The chemical formula for methane is CH_4. Each methane molecule has:

carbon hydrogen

$$CH_4$$

one   four

- one carbon atom (no number in the formula)
- four hydrogen atoms (4 in the formula)
- atoms of two elements joined together
- (1 + 4) = 5 atoms in total.

Remember: you will not find the formulae for any compounds in the periodic table.

Worked example

(a) Give the meaning of the term **ion**.　　(2 marks)

An ion is a charged particle formed when an atom, or group of atoms, loses or gains electrons.

(b) An ion is represented as Na^+.
Explain what this shows.　　(2 marks)

The ion has one positive charge, and it is formed from a sodium atom.

(c) An ion is represented as SO_4^{2-}.
Explain what this shows.　　(2 marks)

The ion has two negative charges, and it is formed from one sulfur atom and four oxygen atoms chemically joined together.

(d) Write the formula for sodium sulfate.　　(1 mark)

Na_2SO_4

In general:
- metal atoms lose electrons to form positive ions
- non-metal atoms gain electrons to form negative ions.

Look at page 5 for help with the atomic structure. For a reminder about how ions form, look at page 10.

Write the charge as a superscript at the top right of the symbol.

You should show a single positive charge as + and a single negative charge as −.

A sodium ion has a single positive charge and a sulfate ion has two negative charges. Two sodium ions balance the charge on one sulfate ion.

Now try this

Use the periodic table on page 8 to help you answer these questions.

1 The formula for bromine is Br_2.
Explain what this shows.　　(2 marks)

2 In copper carbonate, there is one carbon atom and three oxygen atoms for every copper atom. State the formula for copper carbonate.　　(1 mark)

3 The formula for magnesium hydroxide is $Mg(OH)_2$.
Explain what this tells you about the number and type of each atom or ion present.　　(2 marks)

Multiply the numbers inside the brackets by 2.

Equations

Chemical **equations** are used to model the changes that happen in chemical reactions.

Word equations

In a chemical reaction:
- **reactants** are the substances that undergo a chemical change in a reaction
- **products** are the new substances formed.

In general: reactants → products

Two or more reactants or products are separated by a + sign. In a **word equation**, you write the name of each substance, not its formula.

An example word equation

Iron(III) oxide reacts with carbon to form iron and carbon monoxide:

iron(III) oxide + carbon → iron + carbon monoxide

reactants on the left   products on the right

Write it all on one line if possible. If you are going to run out of space, write words below others as shown here.

Balanced equations

- All substances are shown by their formulae.
- The numbers of atoms of each element in the reactants and products are the same.
- You may need to write a number in front of a reactant or product to balance an equation. For example, hydrogen reacts with chlorine to form hydrogen chloride.

1 Write all the symbols and formulae: $H_2 + Cl_2 \rightarrow HCl$

2 Count the atoms of each element on each side, and write numbers if needed:

$$H_2 + Cl_2 \rightarrow 2HCl$$

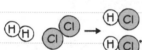

2HCl in the equation means two HCl molecules.

An example balanced equation

The balanced chemical equation for the reaction between iron(III) oxide and carbon is:

$$Fe_2O_3 + 3C \rightarrow 2Fe + 3CO$$

It shows that one unit of iron(III) oxide reacts with three carbon atoms. The products are two iron atoms and three carbon monoxide molecules:

Element	On left	On right
Fe	2	2
O	3	3
C	3	3

State symbols

State symbols show the physical state of each substance in a balanced equation:
- ✓ solid (s)
- ✓ liquid (l)
- ✓ gas (g)
- ✓ aqueous solution (aq) — dissolved in water

Worked example

Aluminium reacts with hot dilute sulfuric acid, H_2SO_4. Aluminium sulfate, $Al_2(SO_4)_3$, and hydrogen, H_2, are formed.
Write a balanced equation for this reaction and include state symbols. **(2 marks)**

$$2Al(s) + 3H_2SO_4(aq) \rightarrow Al_2(SO_4)_3(aq) + 3H_2(g)$$

The unbalanced equation is:
$$Al + H_2SO_4 \rightarrow Al_2(SO_4)_3 + H_2$$
- There are two Al atoms on the right, so a **2** is placed in front of the Al on the left.
- There are three units of SO_4 on the right, so a **3** is placed in front of the H_2SO_4 on the left.
- There are now 3 × H_2 on the left, so a **3** is placed in front of the H_2 on the right.

Now try this

1 Balance these chemical equations:
   (a) $Mg + O_2 \rightarrow MgO$ **(1 mark)**
   (b) $N_2 + H_2 \rightarrow NH_3$ **(1 mark)**
   (c) $CH_4 + O_2 \rightarrow CO_2 + H_2O$ **(1 mark)**
2 Balance these chemical equations and include state symbols.
   (a) $CuO + HNO_3 \rightarrow Cu(NO_3)_2 + H_2O$ (the reaction between copper oxide powder and dilute nitric acid, forming copper nitrate solution and water). **(2 marks)**
   (b) $Fe + Cl_2 \rightarrow FeCl_3$ (the reaction between iron filings and chlorine gas, forming iron(III) chloride powder). **(2 marks)**

Ionic equations

Ionic equations are used to model chemical changes involving ions.

Ions

An **ion** is an electrically charged particle. It is formed when an atom, or group of atoms, loses or gains electrons.

In general:

- hydrogen atoms and metal atoms lose electrons to form positively charged ions
- non-metal atoms gain electrons to form negatively charged ions.

You can revise how ions form on page 10.

Representing ions

You show the charge on an ion using a superscript + or −. For example:

Positive ion	Negative ion
H^+	Cl^-
Na^+	O^{2-}
Mg^{2+}	NO_3^-
Al^{3+}	CO_3^{2-}
NH_4^+	SO_4^{2-}

one S atom, four O atoms, two negative charges

Writing an ionic equation

Oppositely charged ions in solution may join to form an **insoluble** solid. This solid is called a **precipitate**. For example, silver ions and chloride ions in solution form solid silver chloride:

$$Ag^+(aq) + Cl^-(aq) \rightarrow AgCl(s)$$

In an ionic equation:

- all substances are shown by their formula
- the numbers of atoms of each element in the reactants and products are the same
- the total electrical charge is the same on both sides of the equation.

You may need to write a number in front of an ion or substance to balance an ionic equation.

Balancing an ionic equation

Aluminium ions react with hydroxide ions to form aluminium hydroxide.

1   Write all the symbols and formulae:
$$Al^{3+} + OH^- \rightarrow Al(OH)_3$$

2   Count the atoms of each element on each side, and write numbers if needed: $Al^{3+} + 3OH^- \rightarrow Al(OH)_3$

🖩 Maths skills   Using ratios

Simple **ratios** are used to balance equations. The ratio of the charges on the ions (Al^{3+} and OH^-) is 3:1. The ratio of the number of each ion (Al^{3+} and OH^-) in the product is 1:3.

Worked example

Magnesium reacts with copper sulfate solution to form copper and magnesium sulfate solution.
(a) Write a balanced equation for this reaction.
                 (1 mark)

$Mg + CuSO_4 \rightarrow Cu + MgSO_4$

(b) Write a balanced ionic equation for this reaction.
                 (2 marks)

$Mg + Cu^{2+} \rightarrow Cu + Mg^{2+}$

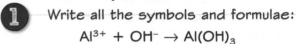

You might be asked to include state symbols:
$$Mg(s) + CuSO_4(aq) \rightarrow Cu(s) + MgSO_4(aq)$$

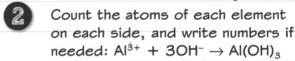

If you write all the ions involved, you get:
$Mg + Cu^{2+} + SO_4^{2-} \rightarrow Cu + Mg^{2+} + SO_4^{2-}$
The sulfate ions, SO_4^{2-}, are unchanged in the reaction. They are called **spectator ions**.
You can leave these out of an ionic equation.

Now try this

1   Lead(II) ions, Pb^{2+}, react with bromide ions, Br^-. Cream-coloured lead(II) bromide forms, $PbBr_2$. Write an ionic equation for the reaction. **(1 mark)**

2   Barium chloride solution, $BaCl_2(aq)$, reacts with sodium sulfate solution, $Na_2SO_4(aq)$.

Solid barium sulfate, $BaSO_4(s)$, and sodium chloride solution, $NaCl(aq)$, form.
(a) Write the balanced equation.
                 (1 mark)

(b) Write an ionic equation for this reaction.
                 (1 mark)

Hazards, risks and precautions

You should be able to evaluate the risks in a practical procedure. You should also be able to suggest suitable precautions.

Hazards

A **hazard** is something that could cause:

- damage or harm to someone or something
- adverse health effects, which may occur immediately or later on.

For example, ethanol is flammable. This is a hazard. If the ethanol ignited, it could cause burns or a fire.

 Practical skills **Risks**

A **risk** is the chance that someone or something will be harmed if exposed to a hazard. The amount of risk depends on factors such as:

✓ how much someone is exposed to a hazard

✓ the way in which exposure happens

✓ how serious the effects of exposure are.

The risk from heating ethanol using a hot water bath is less than when using a Bunsen burner.

Hazard symbols

The labels on containers of hazardous substances include **hazard symbols**. These are intended to:

- warn about the dangers associated with the substance in the container
- let people know about the precautions to take when they use the substance.

Some common hazard symbols

 Harmful or irritant   Flammable   Respiratory sensitiser

 Toxic   Corrosive   Oxidising

Precautions

A **precaution** is something that you can do to reduce the risk of harm from a hazard. Precautions include:

- using a less hazardous substance
- using protective clothing, such as gloves and eye protection
- using a different method or apparatus.

Worked example

A student is preparing a dry sample of copper sulfate. She heats some copper sulfate solution in an evaporating basin. She then allows it to cool. Crystals of copper sulfate appear.

Describe and explain one safety precaution she should use.

(3 marks)

She should heat the solution gently. This reduces the risk that it will spit out of the evaporating basin. The hot solution could cause skin burns or eye damage.

↑

The answer is specific to this activity.

It is not a general lab rule such as not running or not drinking the solution.

Other suitable precautions that could be mentioned, if linked to the activity, include:

- wearing gloves if toxic substances are used
- tying hair back or tucking in a tie if a Bunsen burner is used for heating.

Now try this

1 State **one** reason why hazard symbols are used. **(1 mark)**
2 A student carries out electrolysis on a concentrated sodium chloride solution. Toxic chlorine gas and flammable hydrogen gas are produced.
   Describe **two** precautions the student could take to reduce the risk of harm in this experiment. **(2 marks)**

Atomic structure

Dalton's simple model has changed over time because of the discovery of subatomic particles.

A brief atomic timeline

When	Who	What	
1803	Dalton	Solid atom model: all atoms of an element are identical; different elements have different atoms	
1897	Thomson	Discovers the electron	
1904	Thomson	Plum pudding model: atoms are spheres of positive charge with negative electrons dotted around inside	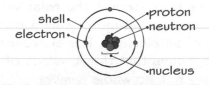
1911	Rutherford	Solar system model: atoms have a positive nucleus surrounded by negative electrons in orbits	
1913	Bohr	Electron shell model: electrons occupy shells or energy levels around the nucleus	
1918	Rutherford	Discovers the proton	
1932	Chadwick	Discovers the neutron	

Atomic structure

An **atom** consists of a central **nucleus**, which:
- contains **protons** and **neutrons**
- is surrounded by electrons in shells.

shell
electron
proton
neutron
nucleus

🖩 Maths skills Relative values

Atoms and their subatomic particles are very small. Their diameters, masses and electrical charges are expressed as **relative values**. For example, the diameter of a hydrogen nucleus is 1.75×10^{-15} m. The diameter of a hydrogen atom is 1.06×10^{-10} m.

The diameter of the nucleus relative to the atom:

$$= \frac{1.75 \times 10^{-15} \text{ m}}{1.06 \times 10^{-10} \text{ m}} = \frac{1}{60\,600}$$

The nucleus is very small compared with the overall size of the atom.

Worked example

Complete the table with the names and properties of subatomic particles. **(3 marks)**

Particle	Relative charge	Relative mass
proton	+1	1
neutron	0	1
electron	−1	$\frac{1}{1836}$

- Atoms have equal numbers of protons and electrons. They have equal numbers of positive and negative charges, and so are neutral overall.
- Electrons have very little mass compared with protons and neutrons. Most of the mass of an atom is concentrated in its nucleus.

Now try this

1. Describe the structure of an atom. **(4 marks)**

2. Explain why atoms are electrically neutral overall. **(2 marks)**

3. The mass of a proton is 1.6726×10^{-27} kg. The mass of an electron is 9.1094×10^{-31} kg. Calculate the mass of an electron relative to a proton.
Give your answer to 4 significant figures. **(2 marks)**

Isotopes

The atoms of an element have identical chemical properties, but can exist as different isotopes.

Numbers of particles

Each atom can be described by its:
- **mass number**, the total number of protons and neutrons in the nucleus
- **atomic number**, the number of protons in the nucleus.

These are in full chemical symbols:

mass number → $^{23}_{11}\text{Na}$ ← atomic number

Atoms of a given element have the same number of protons in the nucleus:
- they have the same atomic number
- this number is unique to that element.

🖩 Maths skills Calculating particle numbers

Use the atomic number and mass number to calculate the number of subatomic particles in an atom.

For example, for sodium, Na:

1   atomic number = 11

number of protons = 11

number of electrons = 11

equal numbers of protons and electrons

2   mass number = 23

neutrons = mass number − atomic number

= 23 − 11 = 12

Isotopes

Isotopes are atoms of an element with:
- the same number of protons
- different numbers of neutrons.

You can recognise isotopes of the same element because they will have the same atomic number, but different mass numbers.

Isotopes of an element have the same chemical properties because they have the same number of electrons.

You can revise electronic configurations on page 9.

Relative atomic mass

Take care not to confuse **relative atomic mass**, A_r, with mass number:

✓ A_r is the mean mass of the atoms of an element, relative to 1/12th the mass of a ^{12}C atom.

A_r values take into account the **relative abundance** or percentage of each isotope in a sample of an element. The existence of isotopes means that the A_r values of elements may not be whole numbers.

Worked example

Gallium has two isotopes: $^{69}_{31}\text{Ga}$ and $^{71}_{31}\text{Ga}$ (sometimes written as gallium-69 and gallium-71). The relative abundance of $^{69}_{31}\text{Ga}$ is 60%. Calculate, to 1 decimal place, the relative atomic mass of gallium. **(3 marks)**

relative abundance of $^{71}_{31}\text{Ga}$ = 100% − 60%

= 40%

relative atomic mass = $\dfrac{(69 \times 60) + (71 \times 40)}{100}$

= $\dfrac{4140 + 2840}{100}$

= 69.8

Multiply the mass number of each isotope by its **relative abundance**, and then add them all together and divide by 100.

Now try this

 1 Bromine has two natural isotopes, $^{79}_{35}\text{Br}$ and $^{81}_{35}\text{Br}$. Explain, in terms of the numbers of subatomic particles, why these are isotopes of the same element. **(4 marks)**

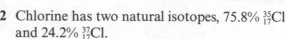 2 Chlorine has two natural isotopes, 75.8% $^{35}_{17}\text{Cl}$ and 24.2% $^{37}_{17}\text{Cl}$.
Calculate the relative atomic mass, A_r, of chlorine.
Give your answer to 1 decimal place. **(2 marks)**

Mendeleev's table

Dmitri Mendeleev's **periodic table** was successful and developed into the modern periodic table. In 1869, Mendeleev arranged all the elements known at the time into a table:

| Mendeleev put the elements in order of the relative atomic masses. |
| He swapped the places of some elements so that elements with similar properties lined up. |
| When these elements were discovered, Mendeleev's predictions fitted the properties very well. |

| He checked the properties of the elements and their compounds. |
| He left gaps where he thought there were other elements, and predicted their properties. |

Pair reversals

Mendeleev thought he had arranged elements in order of increasing relative atomic mass. This was not always true – the positions of elements in some pairs were reversed. For example, Mendeleev put tellurium in group 6 and iodine in group 7. This matched the properties of the elements and their compounds.

However, according to their relative atomic masses, iodine should be first:

A_r of Te = 128 A_r of I = 127

Iodine naturally exists only as ^{127}I. ^{126}Te exists, but ^{128}Te and ^{130}Te are its most abundant isotopes. Their high relative abundance gives tellurium a greater A_r than iodine.

Group 6 and 7 elements

Group 6 elements	Group 7 elements
do not react with water	all react with water
all react with oxygen (except oxygen itself)	do not react with oxygen
all form compounds with hydrogen with the general formula: H_2X	all form compounds with hydrogen with the general formula: HX
6 electrons in their atom's outer shell	7 electrons in their atom's outer shell

This information was not available in Mendeleev's time. See page 9 to revise electronic configurations.

Worked example

Mendeleev published another table in 1871. Part of this is shown on the right. Mendeleev left gaps in his table, shown as * here.

Explain the importance of doing this.   **(3 marks)**

The gaps were for elements not discovered then. Mendeleev used his table to predict the properties of these elements. When they were discovered later, their properties closely matched his predicted properties. This supported the ideas behind his table.

One of the gaps was for 'eka-silicon'. This was discovered in 1886 and named germanium.

| | | Group | | | | | |
1	2	3	4	5	6	7
H						
Li	Be	B	C	N	O	F
Na	Mg	Al	Si	P	S	Cl
K / Cu	Ca / Zn	*	Ti / *	V / As	Cr / Se	Mn / Br
Rb / Ag	Sr / Cd	Y / In	Zr / Sn	Nb / Sb	Mo / Te	* / I

Now try this

1 Describe how Mendeleev arranged the elements known to him.   **(2 marks)**

2 Suggest **one** reason why other scientists at the time thought that Mendeleev's table was not correct.   **(1 mark)**

The periodic table

The modern periodic table is useful for describing and predicting properties of elements.

Atomic number

In Mendeleev's periodic table, atomic number was just the position of an element in the table.

Later discoveries showed that:
- atomic number is actually the number of protons in the nucleus of an atom
- each element has a unique atomic number
- if the elements are arranged in order of increasing atomic number, Mendeleev's pair reversals are explained.

Explaining pair reversals

Iodine should be placed before tellurium according to its relative atomic mass, but after tellurium using its atomic number:

Relative atomic mass	128	127
Element symbol	Te	I
Atomic number	52	53

Mendeleev did not know about atomic structure. He could explain the pair reversal only in terms of the elements' properties.

Features of the modern periodic table

The elements in the modern periodic table are arranged in order of their atomic numbers.

Group 1 2   3 4 5 6 7 0

Period 1

relative atomic mass → 1 H 1 ← atomic number

He 2

The horizontal rows are called **periods**.

| | 7 Li 3 | 9 Be 4 | | | | | | | | | | 11 B 5 | 12 C 6 | 14 N 7 | 16 O 8 | 19 F 9 | 20 Ne 10 |

| | 23 Na 11 | 24 Mg 12 | | | | | | | | | | 27 Al 13 | 28 Si 14 | 31 P 15 | 32 S 16 | 35.5 Cl 17 | 40 Ar 18 |

Metals are on the left-hand side and in the centre.

| | 39 K 19 | 40 Ca 20 | 45 Sc 21 | 48 Ti 22 | 51 V 23 | 52 Cr 24 | 55 Mn 25 | 56 Fe 26 | 59 Co 27 | 59 Ni 28 | 63.5 Cu 29 | 65 Zn 30 | 70 Ga 31 | 73 Ge 32 | 75 As 33 | 79 Se 34 | 80 Br 35 | 84 Kr 36 |

| | 85 Rb 37 | 88 Sr 38 | 89 Y 39 | 91 Zr 40 | 93 Nb 41 | 96 Mo 42 | (98) Tc 43 | 101 Ru 44 | 103 Rh 45 | 106 Pd 46 | 108 Ag 47 | 112 Cd 48 | 115 In 49 | 119 Sn 50 | 122 Sb 51 | 128 Te 52 | 127 I 53 | 131 Xe 54 |

| | 133 Cs 55 | 137 Ba 56 | 139 La 57 | 178 Hf 72 | 181 Ta 73 | 184 W 74 | 186 Re 75 | 190 Os 76 | 192 Ir 77 | 195 Pt 78 | 197 Au 79 | 201 Hg 80 | 204 Tl 81 | 207 Pb 82 | 209 Bi 83 | (209) Po 84 | (210) At 85 | (222) Rn 86 |

Elements with similar properties are placed in the same vertical **groups**.

| | (223) Fr 87 | (226) Ra 88 | (227) Ac 89 | (261) Rf 104 | (262) Db 105 | (266) Sg 106 | (264) Bh 107 | (277) Hs 108 | (268) Mt 109 | (271) Ds 110 | (272) Rg 111 |

Non-metals are on the right-hand side.

Worked example

The table shows information about cobalt and nickel. Their positions in the periodic table were difficult to determine in Mendeleev's time.

Explain how knowledge of atomic structure helps to determine the positions of cobalt and nickel in the periodic table. **(2 marks)**

Relative atomic mass	58.9	58.7
Element symbol	Co	Ni
Atomic number	27	28

Elements are arranged in order of increasing atomic number (number of protons) in the modern table. The atomic number of cobalt is lower than that of nickel.

The answer uses the modern definition of the term **atomic number.** This is the number of protons in the nucleus of an atom. Mendeleev used a different definition.

Now try this

1 The relative atomic mass of an element is often given as a whole number.
   Suggest why cobalt and nickel may **not** appear to be a pair reversal.
   Use the information in the Worked example.

   (1 mark)

Electronic configurations

An **electronic configuration** describes the arrangement of electrons in shells in an atom or ion.

Modelling the arrangement of electrons

In an atom:
- electrons occupy electron **shells**
- shells are filled, starting with the innermost shell
- different shells hold different maximum numbers of electrons.

The diagram shows the electronic configuration of sodium, which has 11 electrons in its atoms.

You can also show it in writing as 2.8.1 (each dot separates two occupied shells).

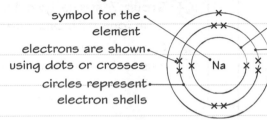

symbol for the element

electrons are shown using dots or crosses

circles represent electron shells

• 1st (inner) shell holds 2 electrons
• 2nd shell holds 8 electrons
• 3rd shell holds 8 electrons (there is only 1 shown here, because sodium atoms only have 11 electrons)

Electronic configurations and the periodic table

The electronic configuration of hydrogen is 1. Below is a 'short form' of the periodic table for the remaining first 20 elements (He to Ca). In general, for these 20 elements:
- metal atoms have up to two electrons in their outer shell (the exception is Al which has three)
- non-metal atoms (except for group O) have four or more electrons in their outer shell.

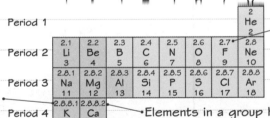

The number of occupied shells is the same as the period number.

The number of electrons in the outer shell is the same as the group number (except for elements in group O, which have full outer shells).

Elements in a group have the same number of electrons in their outer shell, except for helium in group O.

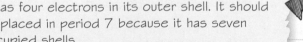

Worked example

Flerovium is an element that was discovered at the end of the twentieth century.

Its electronic configuration is 2.8.18.32.32.18.4. Explain where it should be placed in the periodic table. **(2 marks)**

Flerovium should be placed in group 4 because it has four electrons in its outer shell. It should be placed in period 7 because it has seven occupied shells.

You only need to be able to predict the electronic configurations of the first 20 elements. You should be able to do this in writing (as above) and as diagrams.

The total number of electrons in the atom is 114 (2 + 8 + 18 + 32 + 32 + 18 + 4).

This means that flerovium atoms also have 114 protons. Its atomic number is 114, so flerovium should go between elements 113 and 115.

Now try this

1 Draw the electronic configuration of chlorine, atomic number 17. **(2 marks)**

2 Describe the links between the electronic configuration of an element and its position in the periodic table. **(2 marks)**

Ions

An **ion** is an atom or a group of atoms with a positive or negative charge.

Cations

A **cation** is:

- ✓ a positively charged ion
- ✓ formed when an atom or group of atoms loses one or more electrons.

Cations usually form from hydrogen or metals:

- ✓ Group 1 atoms lose 1 electron to form ions with one positive charge, +.
- ✓ Group 2 atoms lose 2 electrons to form ions with two positive charges, 2+.

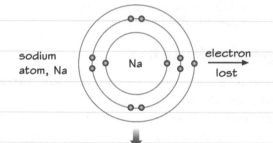

sodium atom, Na   electron lost

sodium ion, Na⁺

Anions

An **anion** is:

- ✓ a negatively charged ion
- ✓ formed when an atom or group of atoms gains one or more electrons.

Anions usually form from non-metals:

- ✓ Group 7 atoms gain 1 electron to form ions with one negative charge, −.
- ✓ Group 6 atoms gain 2 electrons to form ions with two negative charges, 2−.

electron gained → chlorine atom, Cl

chloride ion, Cl⁻

Maths skills **2D models of 3D objects**

In these **dot-and-cross diagrams** each dot or cross represents an electron.

Worked example

Calculate the number of protons, neutrons and electrons in a sodium ion, $^{23}_{11}Na^+$.   **(3 marks)**

number of protons = 11

number of neutrons = (23 − 11) = 12

number of electrons = 11 − 1 = 10

Remember that neutral atoms have equal numbers of protons and electrons.

Reduce the number of electrons for positive ions and increase it for negative ions.

For example, for $^{16}_{8}O^{2-}$:

- number of protons = 8
- number of electrons = 8 + 2 = 10

Now try this

In which groups are these elements found?

Look at the periodic table on page 8.

1 Give the formula for the ions formed by atoms of the following elements:
   - (a) lithium   **(1 mark)**
   - (b) magnesium   **(1 mark)**
   - (c) sulfur   **(1 mark)**
   - (d) bromine   **(1 mark)**

2 Describe the meaning of the term **ion**. **(2 marks)**

3 Calculate the number of protons, neutrons and electrons in the following ions:
   - (a) $^{40}_{20}Ca^{2+}$ **(3 marks)**
   - (b) $^{19}_{9}F^-$ **(3 marks)**

Formulae of ionic compounds

You should be able to write the formulae of ionic compounds from the formulae of their ions.

Naming ions

An ion's name depends on the charge, and whether the ion also contains oxygen. **Compound ions** contain atoms of two different elements.

- Positively charged ions formed from hydrogen or metal atoms take the name of the element:

Formula	Name	
H^+	hydrogen	
Li^+	lithium	
Na^+	sodium	group 1
K^+	potassium	
Mg^{2+}	magnesium	
Ca^{2+}	calcium	group 2
Ba^{2+}	barium	
Al^{3+}	aluminium	group 3
Ag^+	silver	
Cu^{2+}	copper	
Zn^{2+}	zinc	transition metals
Fe^{2+}	iron(II)	
Fe^{3+}	iron(III)	
NH_4^+	ammonium	compound ion

- Negatively charged ions formed from single non-metal atoms take the name of the element, but end in **-ide**.
- Negatively charged ions in compounds containing three or more elements, one of which is oxygen, end in **-ate**:

Formula	Name	
F^-	fluoride	
Cl^-	chloride	
Br^-	bromide	group 7
I^-	iodide	
O^{2-}	oxide	group 6
S^{2-}	sulfide	
NO_3^-	nitrate	
CO_3^{2-}	carbonate	
SO_4^{2-}	sulfate	compound ions
OH^-	hydroxide	

OH^- does not follow the -ate rule.

Worked example

Write down the formulae of these ionic compounds:

(a) sodium fluoride **(1 mark)**

NaF

(b) magnesium oxide **(1 mark)**

MgO

(c) iron(III) oxide **(1 mark)**

Fe_2O_3

(d) sodium nitrate **(1 mark)**

$NaNO_3$

(e) barium nitrate. **(1 mark)**

$Ba(NO_3)_2$

If a compound ion is present **and** more than one of these ions is needed in the formula:
- put the compound ion in brackets, with the number after the closing bracket.

🖩 Maths skills — Balancing charges

This is simple if both ions have:

✓ 1 charge, as in NaF (Na^+ and F^-), or

✓ 2 charges, as in MgO (Mg^{2+} and O^{2-}).

If the ions have different numbers of charges, it may help to multiply them together. For example, for Fe^{3+} and O^{2-}, $(3 \times 2) = 6$. This means that you need:

✓ 2 Fe^{3+} ions to get 6 positive charges,

✓ 3 O^{2-} ions to get 6 negative charges.

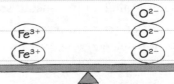

So the formula of iron(III) oxide is Fe_2O_3.

Now try this

You can use the tables on this page to help you.

1 Give the formulae of these ionic compounds:
   (a) calcium sulfide **(1 mark)**
   (b) iron(II) chloride **(1 mark)**
   (c) ammonium hydroxide **(1 mark)**
   (d) ammonium carbonate **(1 mark)**
   (e) sodium sulfate. **(1 mark)**

Properties of ionic compounds

You can explain physical properties of ionic compounds in terms of bonding and structure.

Bonding

Ionic bonds are strong **electrostatic forces** of attraction between oppositely charged ions. For example, when sodium reacts with chlorine to form sodium chloride, NaCl:
- electrons transfer from sodium atoms to chlorine atoms
- Na^+ ions and Cl^- ions form
- Na^+ and Cl^- ions attract each other.

Structure

The ions in an ionic compound form a **lattice** structure which has:
- a regular arrangement of ions
- ionic bonds between oppositely charged ions.

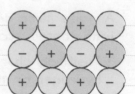

Melting and boiling points

Ionic compounds usually have:
- high melting points
- high boiling points.

As a result, they are in the **solid** state at room temperature. You can explain this in terms of bonding and structure:
- there are many strong ionic bonds
- large amounts of energy must be transferred to the lattice structure to break these bonds.

Solubility in water

Ionic compounds are often **soluble** in water. They **dissolve** to form **aqueous solutions**.

You can revise more about this on page 40.

🖩 Maths skills 3D models

The diagram above models an ionic lattice in two dimensions. However, the lattice extends in three dimensions. It is called a **giant** lattice because it involves very many ions.

The structure of sodium chloride can be modelled in three dimensions as follows:

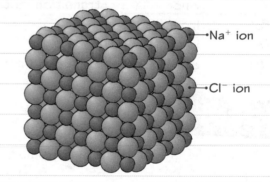

→ Na^+ ion

→ Cl^- ion

Worked example

(a) State why sodium chloride does **not** conduct electricity when it is in the solid state. **(1 mark)**

The ions are not free to move around in a solid.

(b) Explain why sodium chloride **does** conduct electricity when it is molten or in aqueous solution. **(2 marks)**

The ions are free to move around when sodium chloride is liquid or when it is dissolved in water. This means they can carry electric charge from place to place.

Although ions are electrically charged, they are held in fixed positions in the lattice structure.

An electric current is a flow of charge. A substance will conduct electricity if:
- it contains charge carriers (such as ions).

These charge carriers are free to move through the substance.

Now try this

1 Aluminium oxide is an insoluble ionic compound.
   (a) State why solid aluminium oxide cannot conduct electricity. **(1 mark)**
   (b) Describe how you could make aluminium oxide conduct electricity. **(2 marks)**
   (c) Suggest why the method described in (b) may be expensive. **(2 marks)**

Covalent bonds

A **covalent bond** is formed when a pair of electrons is shared between two atoms.

A shared pair of electrons

Covalent bonds:
- are strong
- form between non-metal atoms
- often produce **molecules**, which can be elements or compounds.

A hydrogen atom can form one covalent bond. Usually, for atoms of other non-metals:
- number of bonds = (8 – group number).

Group	Example	Covalent bonds
4	carbon, C	4
5	nitrogen, N	3
6	oxygen, O	2
7	chlorine, Cl	1
0	helium, He	none

Helium and other elements in group 0 have full outer shells. They do not transfer or share electrons, so they are unreactive.

Modelling covalent bonds

There are three ways you can represent a covalent bond, e.g. hydrogen, H_2:

1 dot-and-cross (with shells)

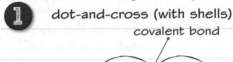
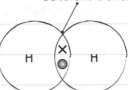

2 dot-and-cross (without shells)

H ● H
   ×

3 **structural formula**

H–H

Maths skills The size of atoms is around 1×10^{-10} m and the size of simple molecules is around 1×10^{-9} m.

Some examples of dot-and-cross diagrams

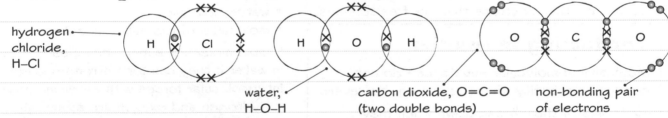

hydrogen chloride, H–Cl

water, H–O–H

carbon dioxide, O=C=O (two double bonds)

non-bonding pair of electrons

When you draw diagrams like these:
- draw overlapping circles to represent the outer shells
- draw a dot to represent an electron from one atom, and a cross to represent an electron from the other atom (make sure each bonding pair of electrons is between the two atoms)
- make sure that each atom (apart from hydrogen) has eight electrons in its outer shell.

Worked example

Methane, CH_4, can be represented by this structure:
Draw a dot-and-cross diagram to represent methane. **(1 mark)**

H
|
H—C—H
|
H

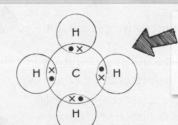

Hydrogen atoms do not have non-bonding pairs of electrons, but other atoms may do.

Now try this

The molecule contains a double bond.

1 Describe covalent bonding. **(2 marks)**

2 Draw a dot-and-cross diagram to represent a molecule of oxygen, O_2. **(2 marks)**

Simple molecular substances

You can explain the physical properties of simple molecular substances in terms of their bonding and structure.

Bonding

A **simple molecule** consists of just a few atoms, joined to each other by strong covalent bonds. Simple molecular substances can be:

- non-metal elements, such as H_2, O_2, Cl_2
- compounds of non-metals, such as HCl, H_2O, CH_4.

Simple molecular substances usually have:

- low melting points
- low boiling points.

They are usually in the **gas** or **liquid** state at room temperature because of this.

Simple molecular substances with relatively large molecules, such as wax, are in the solid state at room temperature.

Melting and boiling points

There are weak attractive forces between molecules, called **intermolecular forces**.

When a simple molecular substance such as oxygen, O_2 or $O=O$, melts or boils:

- ✓ intermolecular forces are overcome
- ✓ covalent bonds do not break.

strong covalent bonds between atoms

not broken in state changes

weak intermolecular forces between molecules

overcome in state changes

Non-conductors of electricity

Simple molecular substances do not conduct electricity when solid, liquid or gas. This is because their molecules:

- are not electrically charged, and
- do not contain electrons that are free to move.

Conducting in solution

Most simple molecular substances do not conduct electricity when in solution. However:

- ✓ some of them break down when they dissolve in water, forming ions
- ✓ these ions can move around, so the solution conducts electricity.

Solubility in water

Many simple molecular substances are **insoluble** in water. The intermolecular forces between water and these substances are weaker than those between:

- water molecules
- molecules of the substance itself.

Simple molecular substances dissolve in water if they can form strong enough intermolecular forces with water molecules:

- Hydrogen and oxygen are **sparingly soluble** (very little dissolves).
- Chlorine, carbon dioxide, sulfur dioxide and ammonia are soluble in water.
- Ethanol and ethanoic acid are soluble.
- Sugar is soluble in water.

Worked example

Nitrogen is a colourless, unreactive gas at room temperature.
Explain why it is suitable for use as an insulator in high-voltage electrical transformers.   **(2 marks)**

Nitrogen is unreactive. So, it will not react with the materials used in the transformer. Nitrogen can insulate the parts in the transformer because it does not conduct electricity.

A nitrogen molecule consists of two nitrogen atoms joined together by a triple covalent bond: N≡N

A lot of energy must be transferred to break this very strong bond. This is why nitrogen is unreactive.

The answer includes only the properties of nitrogen that make it useful in transformers.

Now try this

1 Ammonia, NH_3, is a simple molecular substance. Explain, in terms of intermolecular forces, why it is in the gas state at room temperature. **(2 marks)**

2 Petrol and water are simple molecular substances. Explain why petrol does not dissolve in water. **(3 marks)**

Giant molecular substances

Giant molecules contain very many atoms, rather than just a few.

Bonding and structure

A **giant molecule** consists of many atoms. In giant molecules, the atoms are:
- joined by strong covalent bonds
- arranged in a regular lattice structure.

Giant molecular substances can be:
- non-metal elements, such as carbon
- compounds such as silica.

Giant molecular substances usually have:
- high melting points
- high boiling points.

They are solids at room temperature. A lot of energy must be transferred to break the many strong covalent bonds during melting and boiling. Giant molecular substances are insoluble in water.

Modelling giant molecules

There are many atoms in an entire giant molecule. You cannot represent an entire giant molecule using displayed formulae or dot-and-cross diagrams.

This is a **ball-and-stick model** of a small part of a silica molecule (found in sand):

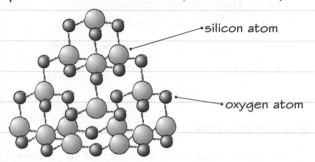

- silicon atom
- oxygen atom

The formula of silica is SiO_2. As with other giant molecular substances, this is its **empirical formula** (see page 20).

Diamond

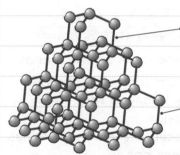

- each atom is bonded to four others
- strong covalent bonds between atoms

Graphite

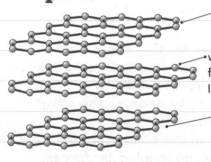

- each atom is bonded to three others
- weak intermolecular forces between layers
- strong covalent bonds between atoms in a layer

Diamond and graphite are both forms of carbon. They are giant molecular substances.

Worked example

Graphite is used to make electrodes because it conducts electricity. Explain why it conducts electricity but diamond does not. **(4 marks)**

A carbon atom can form four covalent bonds. In graphite each carbon atom only forms three covalent bonds. The non-bonding outer electrons become delocalised. This means that they can move through the structure, so graphite conducts electricity. Diamond does not have delocalised electrons and does not conduct.

The rigid lattice structure and strong bonds of diamond make it very hard. This is why it is useful for cutting tools.

In graphite, the weak intermolecular forces let the layers slide over each other. This is why it is slippery and useful as a lubricant.

Metals also have delocalised electrons. This is why metals are good conductors of electricity. You can revise the structure and properties of metals on page 17.

Now try this

Your answer must include at least one similarity and one difference.

 1 Compare and contrast the structure and bonding of diamond and graphite. **(4 marks)**

 2 Explain, in terms of bonding and structure, why graphite is used as a lubricant. **(2 marks)**

15

Other large molecules

Graphene and fullerenes are forms of carbon that exist as giant molecules.

Graphene

Graphene is a giant molecular substance. Its structure resembles a single layer of graphite:

- Each carbon atom is covalently bonded to three other carbon atoms.
- It has a regular lattice structure.

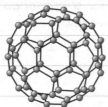 ←interlocking hexagonal rings of carbon atoms

Properties of graphene

Graphene conducts electricity:

- ✓ The non-bonding outer electrons become delocalised.
- ✓ They can move through the structure.

Graphene is very strong and flexible:

- ✓ It contains many strong covalent bonds.

Graphene is almost transparent:

- ✓ Its layers are just one atom thick.

Fullerenes

Fullerenes resemble a sheet of graphene rolled to form:

- hollow balls, often called **buckyballs**

Buckminsterfullerene, C_{60}, has carbon atoms arranged in pentagons as well as hexagons. Materials made from buckyballs:

- conduct electricity because they have delocalised electrons
- are soft when in the solid state because they have weak intermolecular forces.

- hollow tubes, called carbon **nanotubes**.

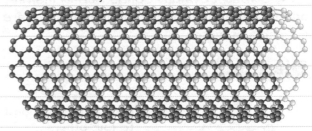

Nanotubes have closed ends or open ends. They can be several mm long. Nanotubes:

- conduct electricity because they have delocalised electrons
- are very strong because the structure has many strong covalent bonds.

Worked example

The diagram shows a section of a molecule of poly(ethene), a simple polymer.

Describe the structure of poly(ethene).   **(2 marks)**

Poly(ethene) consists of large molecules containing chains of carbon atoms. These atoms are joined to each other, and to hydrogen atoms, by covalent bonds.

Polymers are large molecules made from many smaller molecules, called **monomers**, joined together.

Revise addition polymers on page 101 and condensation polymers on page 102.

Poly(ethene) is not a fullerene:

- It is a **hydrocarbon** (a compound of carbon and hydrogen).

Polymer molecules are described as **macromolecules** rather than giant covalent molecules.

Now try this

1 Describe the properties of graphene that make it suitable for use in flexible touch screens.   **(2 marks)**
2 Explain why buckminsterfullerene has a much lower melting point than diamond.   **(3 marks)**

Metals

Most elements are metals, and are placed on the left-hand side of the periodic table.

Metals versus non-metals

Some typical physical properties:

Property	Metals	Non-metals
appearance	shiny	dull
electrical conduction	good conductors	poor conductors
density	high	low
melting point	high	low

Mercury is liquid at room temperature.

Diamond and graphite have very high melting points. Graphite conducts electricity.

In addition:
- metals are **malleable** – they can be pressed into shape without shattering
- non-metals are **brittle** in the solid state – they shatter when bent or hit.

Metallic structure and bonding

A metal:
- ☑ consists of a giant lattice of positively charged metal ions
- ☑ has a 'sea' of delocalised electrons.

The delocalised electrons come from the outer shells of the atoms.

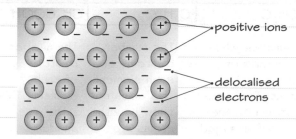

positive ions

delocalised electrons

Metallic bonds are strong electrostatic forces of attraction between positive metal ions and delocalised electrons.

Malleable metals

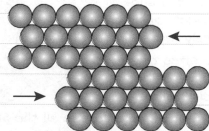

If a force is applied to a metal:
- layers of positive ions slide over each other
- the metal changes shape without shattering.

Insoluble metals

Metals are insoluble in water. However, some metals do seem to dissolve in water.

This is because they react with the water to produce soluble metal hydroxides. These dissolve, exposing more metal to the water.

For example, sodium reacts with water, forming sodium hydroxide solution and hydrogen:

$$2Na(s) + 2H_2O(l) \rightarrow 2NaOH(aq) + H_2(g)$$

You can revise the reactions of sodium and other group 1 metals on page 72.

You can revise the reactions of sodium and other group 1 metals on page 72.

Worked example

Copper is used in electricity cables. Explain, in terms of structure and bonding, why metals are good conductors of electricity.   **(2 marks)**

Metals contain delocalised electrons which can move through the structure of the metal.

Copper is also **ductile**. It can be pulled to make wires without breaking. This is because its layers of positive ions can slide over each other.

The answer makes it clear that the delocalised electrons, not the metal ions, move through the structure.

Now try this

1 Aluminium is used to make overhead mains electricity cables. In terms of its structure and bonding, explain why aluminium is:
   (a) ductile   **(2 marks)**
   (b) a good conductor of electricity.   **(2 marks)**

2 Mercury is a liquid metal at room temperature. State whether it should conduct electricity, giving a reason for your answer.   **(1 mark)**

Limitations of models

The structure and bonding of different substances are represented using **models**. Different models have different features and limitations. All the examples on this page refer to ethanoic acid.

Written formulae

The formula for a substance can be written as:
- an empirical formula
- a molecular formula
- a **structural** formula.

Empirical	Molecular	Structural
CH_2O	$C_2H_4O_2$	CH_3COOH

The simplest whole number ratio of atoms of each element. This does not show how the atoms are arranged, or (usually) the actual number of atoms.

The number of atoms of each element. This does not show how the atoms are arranged.

The number of atoms of each element. This does give an idea of how they are arranged.

Drawn structures

When you draw a structure, you should show all the covalent bonds in the molecule.

This carbon atom is covalently bonded to three hydrogen atoms. It is also covalently bonded to another carbon atom.

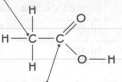

This carbon atom is covalently bonded to two oxygen atoms. It has a double bond to one of them.

This model does not show:
- the molecule's three-dimensional shape
- the bonding and non-bonding electrons.

Ball-and-stick models

You can draw **ball-and-stick models**. You can also make them using plastic modelling kits.

These models show:
- how each atom is bonded to other atoms
- the molecule's three-dimensional shape.

They do not show the bonding and non-bonding electrons, or each element's chemical symbol.

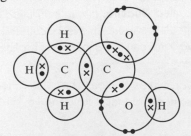

Space-filling models

Similar to ball-and-stick models but more accurately represent:

☑ the sizes of atoms relative to their bonds.

You may not be able to see all the atoms in a complex space-filling model.

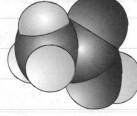

Worked example

The diagram below is a dot-and-cross diagram.

Describe the information it provides, and state a limitation of this model.   **(4 marks)**

Dot-and-cross diagrams are two-dimensional models. This one may lead you to think incorrectly that ethanoic acid molecules are flat.

The diagram shows the symbol for each atom in the molecule. It also shows how each atom is bonded to other atoms. The pairs of electrons in each covalent bond are shown by dots and crosses. Non-bonding pairs of electrons in the outer shells are included.

It does not show the three-dimensional shape of the molecule.

Now try this

1 State **one** limitation of each model described on this page.   **(7 marks)**

Relative formula mass

You should be able to calculate relative formula masses when given relative atomic masses.

Calculating relative formula mass

Relative formula mass has the symbol M_r.

To calculate the M_r of a substance, add together the relative atomic masses of all the atoms shown in its formula:

oxygen molecule – formula O_2
relative atomic mass of oxygen = 16
relative formula mass = 2 × 16
= 32

No units

M_r values are just numbers.

This is because an M_r value is the mass of a molecule or unit of a substance compared with 1/12th the mass of a ^{12}C atom. The M in M_r stands for 'molecular'.

You might see or hear the term 'relative molecular mass'. This really applies only to covalent substances.

Worked example

Calculate the relative formula mass of aluminium oxide, Al_2O_3. **(1 mark)**
(Relative atomic masses: Al = 27, O = 16)
atoms in Al_2O_3:
(2 × Al) + (3 × O)
M_r = (2 × 27) + (3 × 16)
= 54 + 48
= 102

You do not need to learn any relative atomic masses. You will be given them in questions or you can find them on the periodic table.

This answer shows you the working out needed to obtain the answer.

If you show the working for steps in the calculation you may gain some marks even if your final answer is incorrect.

Worked example

Calculate the relative formula mass of calcium nitrate, $Ca(NO_3)_2$. **(1 mark)**
(Relative atomic masses: Ca = 40, N = 14, O = 16)
atoms in $Ca(NO_3)_2$:
(1 × Ca) + (2 × 1 × N) + (2 × 3 × O)
M_r = (1 × 40) + (2 × 14) + (6 × 16)
= 40 + 28 + 96
= 164

Maths skills You may find it easier if you first add up the A_r values for the atoms inside the brackets:
M_r of NO_3 = 14 + (3 × 16)
= 14 + 48
= 62
Then multiply your answer by the number outside, and add that to the remaining A_r values:
M_r of $Ca(NO_3)_2$ = (2 × 62) + 40
= 124 + 40
= 164

Now try this

Relative atomic masses: H = 1, C = 12, O = 16, Na = 23, Al = 27, S = 32, Cl = 35.5, Cu = 63.5

1 Calculate the relative formula masses, M_r, of the following substances.
(a) H_2O **(1 mark)**
(b) CO_2 **(1 mark)**
(c) NaOH **(1 mark)**
(d) CCl_4 **(1 mark)**
(e) $CuCl_2$ **(1 mark)**
(f) Na_2SO_4 **(1 mark)**
(g) $Al(OH)_3$ **(1 mark)**
(h) $Al_2(CO_3)_3$ **(1 mark)**

Empirical formulae

An **empirical formula** is the simplest whole number ratio of atoms of each element in a compound.

Calculating an empirical formula

A 10 g sample of a compound **X** contains 8 g of carbon and 2 g of hydrogen.

		C	H
1	Write the symbol of each element as a header	C	H
2	Write down the mass of each element in g	8	2
3	Write down the A_r of each element	12	1
4	For each element, calculate: mass ÷ A_r	$\frac{8}{12} = 0.667$	$\frac{2}{1} = 2$
5	Divide each answer by the smallest answer (0.667 here)	$\frac{0.667}{0.667} = 1$	$\frac{2}{0.667} = 3$
6	You may then need to multiply all the numbers to remove fractions, then write out the empirical formula	CH_3	

Finding a molecular formula

You can find the molecular formula of a compound from its empirical formula:
- if you know its relative formula mass, M_r.

The M_r of **X** in the example above is 30:

1 Calculate the M_r of the empirical formula:

M_r of $CH_3 = 12 + (3 \times 1) = 15$

2 Divide the M_r of **X** by answer 1:

$\frac{30}{15} = 2$

3 Multiply each number in the empirical formula by answer 2:

CH_3 becomes C_2H_6 – the molecular formula

🧪 Practical skills · Determining empirical formula

You need to be able to describe an experiment to determine an empirical formula.

The apparatus below can be used to obtain results to do this for magnesium oxide.

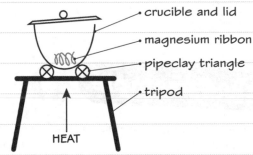

- crucible and lid
- magnesium ribbon
- pipeclay triangle
- tripod

HEAT

The crucible and its contents are weighed before and after heating the magnesium.

Worked example

The table shows the results of an experiment to find the empirical formula of magnesium oxide.

Object	Mass (g)
empty crucible and lid	19.06
crucible, lid and Mg before heating	19.42
crucible, lid and Mg after heating	19.66

(a) Calculate the mass of magnesium used.
(1 mark)

mass of magnesium = 19.42 – 19.06
= 0.36 g

(b) Calculate the mass of oxygen gained.
(1 mark)

mass of oxygen = 19.66 – 19.42
= 0.24 g

Now try this

1 (a) Use the masses given in the Worked example to determine the empirical formula of magnesium oxide. (Relative atomic masses: O = 16, Mg = 24)   **(3 marks)**

(b) In the experiment described on this page, a lid is needed on the crucible. The lid must be kept slightly open during heating. Give **two** reasons why.   **(2 marks)**

Conservation of mass

According to the **law of conservation of mass**, the total mass of reactants and products stays constant during a chemical reaction. The total mass before and after a reaction is the same.

Closed systems

A **closed system** is a situation in which no substances can enter or leave during a reaction.

Closed systems include:
- reactions in a sealed container, such as a flask fitted with a bung
- precipitation reactions in a beaker.

In a **precipitation reaction**, two **soluble** reactants form an **insoluble** product – the **precipitate**. For example:

$NaCl(aq) + AgNO_3(aq) →$ cloudy white
$NaNO_3(aq) + AgCl(s)$ ——•precipitate

forms a

The total mass of the beaker and its contents stays the same during the formation of the silver chloride precipitate.

Non-enclosed systems

A **non-enclosed system** is a situation in which substances can enter or leave during a reaction.

Non-enclosed or 'open' systems include:
- reactions in an open flask, where a substance in the gas state may enter or leave.

Mass is conserved, but you will observe the following:
- The mass of a reactive metal **increases** if it is heated in air. This is because oxygen atoms combine with metal atoms to form a metal oxide.
- The mass of a reactive non-metal or a fuel **decreases** if it is heated in air. This is because products in the gas state escape from the container.
- The mass of a metal carbonate **decreases** if it is heated. This is because carbon dioxide gas is produced and this escapes from the container.

Worked example

0.5 g of lithium metal is added to water. Calculate the mass of hydrogen gas produced.

(4 marks)

(Relative atomic masses: Li = 7, H = 1)

Always start with the balanced equation. You do not need state symbols.

Work out the relative masses and multiply by the balancing numbers. You only need to do this for the substances asked for in the question.

Divide by the number for lithium to find the mass of hydrogen produced for each gram of lithium.

Multiply by the mass of lithium given in the question.

$2Li + 2H_2O → 2LiOH + H_2$
$2 × 7 \quad\quad\quad 1 × (1 + 1)$
$14 \quad\quad\quad\quad\quad 2$
14 g of Li $→$ 2 g of H_2
$1g \quad → \quad \frac{2}{14}$ g $= 0.143g$
0.5 g $→$ 0.143 g $× 0.5$
$\quad\quad\quad = 0.072$
0.072 g of hydrogen is produced.

Now try this

1. Methane, CH_4, burns in oxygen to form carbon dioxide and water.
   $CH_4 + 2O_2 → CO_2 + 2H_2O$
   Calculate the mass of oxygen used when 10 g of methane burns.
   (Relative atomic masses: C = 12, H = 1, O = 16)
   (4 marks)

2. Potassium reacts with chlorine to form potassium chloride, KCl.
   Calculate the mass of potassium needed to produce 20 g of potassium chloride. (Relative atomic masses: K = 39, Cl = 35.5) **(4 marks)**

Reacting mass calculations

A balanced chemical equation is also called a **stoichiometric equation**.

Limiting or in excess?

A reactant is **in excess** if there is enough:
- to react with all the other reactant
- for some of it to be left over when the reaction stops.

The other reactant is the **limiting reactant**. For example, when magnesium reacts with hydrochloric acid the magnesium ribbon gradually becomes smaller:

Situation after reaction	Magnesium	Hydrochloric acid
some magnesium left	in excess	limiting
no magnesium left	limiting	in excess

The **stoichiometry** of a reaction is to do with the ratio of the amounts of reactants and products involved.

When you balance a chemical equation you are finding the 'stoichiometric equation'.

Limiting reactant and mass of product

The mass of product formed is controlled by the mass of the reactant that is not in excess:

✓ the reaction continues until all the particles of the limiting reactant have been used up.

Maths skills   Directly proportional

The mass of product is directly proportional to the mass of the limiting reactant, for example:

positive gradient **and** passes through the origin

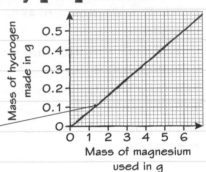

Worked example

An oxide of copper is heated with excess hydrogen, H_2. In the reaction, 1.27 g of copper and 0.18 g of water, H_2O, form.
Use this information to determine the stoichiometry of the reaction. (Relative atomic masses: H = 1, O = 16, Cu = 63.5)   **(4 marks)**

amount of Cu = 1.27/63.5 = 0.02
M_r of H_2O = 1 + 1 + 16 = 18
amount of H_2O = 0.18/18 = 0.01
ratio of Cu:H_2O = 0.02:0.01
              = 2:1
right-hand side must be → 2Cu + H_2O
so the left-hand side must be Cu_2O + H_2 →
equation is: Cu_2O + H_2 → 2Cu + H_2O

Hydrogen is in excess, so the oxide of copper must be the limiting reactant. This means that the amount of the oxide of copper controls the amounts of copper and water produced.

The calculated amounts of each product are in moles (mol). You can revise mole calculations on page 24.

The oxide of copper must contain two copper atoms for every oxygen atom. This is why its formula is Cu_2O (and not CuO, the black copper(II) oxide you normally see).

Now try this

1 Explain why, in a chemical reaction, the mass of the reactant not in excess determines the mass of product formed.   **(2 marks)**

2 Powdered carbon is heated with excess steam, producing 3.3 g of carbon dioxide and 0.30 g of hydrogen. Determine the stoichiometry of the reaction. (Relative formula masses: CO_2 = 44, H_2 = 2)   **(4 marks)**

Concentration of solution

You need to be able to calculate the concentration of solutions in g dm⁻³.

Solute, solvent and solution

A **solution** is a mixture of a solute in a solvent:
- the **solute** is the substance that dissolves
- the **solvent** is the substance that the solute dissolves in.

Water is the solvent in an **aqueous solution**. The state symbol for an aqueous solution in balanced equations is (aq). The symbol (l) is for substances in the liquid state.

Mass and volume

To calculate the **concentration** of a solution, you need to know:
- the mass of solute in **grams**, g, and
- the volume of solution in **cubic decimetres**, dm³: —•If you are making a solution, you can use the volume of the solvent instead.

dm³ and cm³

Measuring cylinders and other lab apparatus show volumes in cubic centimetres, cm³. You need to convert these measurements into cubic decimetres, dm³, when you calculate concentrations. It helps to know that:
- ✓ 1 dm³ = 10 × 10 × 10 = 1000 cm³
- ✓ to convert cm³ to dm³, divide by 1000.

Concentration in mol dm⁻³

You can revise calculating the concentration of a solution in moles per cubic decimetre, mol dm⁻³, on page 60.

Mass, volume and concentration

You use this equation to calculate the concentration of a solution in g dm⁻³:

$$\text{concentration (g dm}^{-3}\text{)} = \frac{\text{mass of solute (g)}}{\text{volume of solution (dm}^3\text{)}}$$

LEARN IT! IT'S NOT ON THE EQUATIONS LIST

 Units

The unit g dm⁻³ means 'grams per cubic decimetre'. You may also see it written as g/dm³.

Maths skills Rearranging equations

You need to be able to change the subject of an equation. For example:
- ✓ mass of solute = concentration × volume
- ✓ volume = $\dfrac{\text{mass of solute}}{\text{concentration}}$

Worked example

2.50 g of sodium hydroxide is dissolved in 250 cm³ of water. Calculate the concentration of the solution formed in g dm⁻³. **(2 marks)**

$$250 \text{ cm}^3 = \frac{250}{1000} = 0.250 \text{ dm}^3$$

$$\text{concentration} = \frac{2.50 \text{ g}}{0.250 \text{ dm}^3} = 10 \text{ g dm}^{-3}$$

Remember to convert the volume to dm³ if it is given to you in cm³.

The units are shown in the concentration calculation here. This makes it easier for you to see how it is done. You do not need to show units in your working out, but you must show the units in your final answer.

Now try this

1 Calculate the concentrations of the following solutions formed:
  (a) 0.40 g of glucose dissolved in 0.50 dm³ of water. **(1 mark)**

  (b) 1.25 g of copper chloride dissolved in 100 cm³ of water. **(2 marks)**

2 Calculate the mass of sodium hydroxide needed to make 150 cm³ of a 40 g dm⁻³ solution. **(2 marks)**

Avogadro's constant and moles

You need to be able to carry out calculations involving Avogadro's constant and the mole.

The mole

In chemistry, the 'amount' of a substance does not refer to its volume or mass.

The **mole** is the unit for amount of substance. It's shown as **mol** in calculations and values. One mole (1 mol) of particles of a substance is:

- Avogadro's constant number of particles (atoms, ions or molecules) of that substance.

The mass of 1 mol of particles is the 'relative particle mass' in grams.

> **Maths skills — Avogadro's constant**
>
> **Avogadro's constant** is 6.02×10^{23} mol^{-1} (to three significant figures).
>
> The number 6.02×10^{23} is a number in its **standard form**. In general, you write such numbers as:
>
> $1 \leq a < 10$ (a is between 1 and 10) ⟶ $a \times 10^n$ ⟵ an integer (whole number)
>
> 6×10^{23} means 6 followed by 23 zeros:
> 600 000 000 000 000 000 000 000

Masses from moles

The mass of 1 mol of a substance is its A_r or M_r in grams. The table shows you three examples.

Substance	Mass of 1 mol
carbon, C, $A_r = 12$	12 g
oxygen, O_2, $M_r = 32$	32 g
carbon dioxide, CO_2, $M_r = 44$	44 g

To calculate the mass of an amount of substance:

A_r for atoms ⟶ mass (g) $= M_r \times$ amount (mol)

Moles from masses

To calculate the amount of substance from a given mass:

$$\text{amount (mol)} = \frac{\text{mass (g)}}{A_r \text{ or } M_r}$$

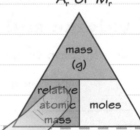

Worked example

(a) Calculate the amount, in mol, of ethane molecules in 45 g of ethane, C_2H_6. (Relative atomic masses: C = 12, H = 1) **(2 marks)**

M_r of $C_2H_6 = (2 \times 12) + (6 \times 1) = 30$

amount $= \dfrac{45}{30} = 1.5$ mol

(b) Calculate the number of ethane molecules in 45 g of ethane. Give your answer to two significant figures. **(1 mark)**

number $= 1.5 \times 6.02 \times 10^{23}$
$\quad\quad\quad = 9.0 \times 10^{23}$ ethane molecules

> You also need to be able to calculate the mass of a substance if you are given its amount. For example, the mass of 2.0 mol of ethane = 30 × 2.0 = 60 g

> Each ethane molecule contains eight atoms (two carbon atoms and six hydrogen atoms). The number of atoms in 45 g of ethane = 8 × 9 × 10^{23} = 7.2 × 10^{24} (to two significant figures)

Now try this

> You will need to use Avogadro's constant and the A_r of carbon.

1 (a) Calculate the amount, in mol, of molecules in 22.5 g of water. (M_r of $H_2O = 18$) **(1 mark)**
   (b) Calculate the amount, in mol, of atoms in 22.5 g of water. **(1 mark)**
   (c) Use your answer to **(b)** to calculate the number of atoms in 22.5 g of water. **(1 mark)**

2 (a) Calculate the number of atoms in 6.0 g of diamond.
      Give your answer to two significant figures. (A_r of C = 12) **(2 marks)**
   (b) Calculate the mass, in grams, of 1.00×10^{12} carbon atoms.
      Give your answer to two significant figures. **(2 marks)**

Extended response – Types of substance

There will be one or more 6-mark question on your exam paper. For these questions, you will need to think scientifically and structure your answer logically, showing how the points you make are related to each other.

You can revise the topics for this question, which is about the **structure and bonding** of different types of substance, on pages 12–17.

Worked example

Sodium chloride is an ionic compound produced in the reaction between sodium and chlorine. The table shows information about how well these three substances conduct electricity at room temperature.

Substance	Ability to conduct
sodium	conducts
chlorine	does not conduct
sodium chloride	does not conduct

Explain, in terms of structure and bonding, the differences in the ability to conduct electricity.

(6 marks)

Sodium is a metal. It consists of a regular lattice of positively charged sodium ions attracted to a sea of delocalised electrons. These electrons are free to move, so solid sodium conducts electricity.

On the other hand, chlorine is a simple molecular substance. Its molecules are attracted to each other by weak intermolecular forces. Even though its molecules are free to move they are uncharged, so chlorine does not conduct.

Sodium chloride is an ionic compound. In the solid state, it contains oppositely charged ions held together strongly in a regular lattice. Therefore, even though it contains charged particles, these are not free to move. This is why solid sodium chloride does not conduct electricity.

Command word: Explain

When you are asked to **explain** something, it is not enough just to state or describe it. Your answer **must** contain some reasoning or justification of the points you make. Your explanation **can** include mathematical explanations, if calculations are needed.

In this example Exam skills question, no calculations are needed.

You should give a description of the structure of a solid metal, making it clear that solid sodium contains sodium ions. You should also give the reason why sodium conducts electricity.

You should use **connectives** as you move from one idea to the next. Here the use of 'on the other hand' shows that this part of the answer is in contrast to the first part of the answer.

This part of the answer includes another connective: 'therefore' indicates that the sentence before has a consequence. In this case the lattice structure prevents the conduction of electricity.

You need to show comprehensive knowledge and understanding using relevant scientific ideas to support your explanations. You should consider each substance in turn, giving clear lines of reasoning. Leave out any information or ideas that are unnecessary to answer the question.

Now try this

Mercury is a metal and paraffin oil is a simple molecular compound. Both exist as liquids at room temperature. Zinc chloride exists as a liquid above 290 °C. Paraffin oil does not conduct electricity but the other two liquids do.

Explain, in terms of structure and bonding, the differences in the ability to conduct electricity.

(6 marks)

States of matter

The **particle theory** models the states of matter, with particles described as hard spheres.

	Solid	Liquid	Gas
Particle diagram	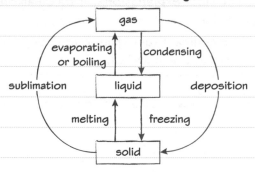		
Arrangement of particles	close together regular pattern	close together random	far apart random
Movement of particles	vibrate about fixed positions	move around each other	fast in all directions
Relative energy of particles	least stored energy	⟶	most stored energy

State changes

The interconversions between states of matter are called **state changes**.

gas

evaporating or boiling — condensing

sublimation — liquid — deposition

melting — freezing

solid

During a state change:
- energy is transferred to or from the particles
- the arrangement of particles changes
- the movement of particles changes.

Melting and boiling points

At its **melting point**, a substance begins to:
- ✓ melt if energy is transferred to the particles
- ✓ freeze if energy is transferred to the surroundings.

At its **boiling point**, a substance begins to:
- ✓ boil if energy is transferred to the particles
- ✓ condense if energy is transferred to the surroundings.

A substance evaporates if it changes from a liquid to a gas below its boiling point:
- ✓ Particles with high enough energy leave the surface of the liquid.
- ✓ The remaining particles have less energy.
- ✓ The liquid cools down unless it is heated.

Worked example

The table shows some data about bromine.

Melting point (°C)	Boiling point (°C)
−7	59

(a) Predict the state of bromine at 25 °C. **(1 mark)**

liquid.

(b) Describe, in terms of the particle theory, what happens when bromine boils. **(3 marks)**

Forces of attraction between bromine particles are overcome and the particles move far apart. The particles no longer just move around each other, but move quickly in all directions.

A substance is a:
- solid below its melting point
- gas above its boiling point
- liquid between its melting and boiling points.

See pages 10–17 to revise the different types of bonds and forces. Some forces of attraction are overcome during melting, and the remainder during boiling.

State changes are **physical changes** because the particles themselves are unchanged.

Chemical reactions cause **chemical changes** – the particles are changed by rearranging atoms.

Now try this

1 Naphthalene is a solid that can change directly from a solid to a gas at room temperature. Name this change of state, and describe what happens to the particles. **(5 marks)**

Pure substances and mixtures

Elements and compounds

An **element** is a substance that consists only of atoms with the same atomic number (the same number of protons in their nucleus), e.g.

- Hydrogen is an element because its atoms all have one proton in their nucleus.
- Oxygen is a different element because its atoms all have eight protons.

A **compound** is a substance that consists of atoms of two or more different elements, chemically joined together.

For example, water is a compound. This is because it consists of hydrogen atoms and oxygen atoms chemically joined together.

You can revise **ionic bonds** on page 10 and **covalent bonds** on page 13.

Atoms and molecules

Elements exist as atoms or molecules.

Argon is in group 0. It has no tendency to gain, lose or share electrons. It exists as single atoms (that is, it is **monatomic**).

Hydrogen and oxygen exist as simple molecules.

Carbon exists as giant molecules (diamond, graphite and graphene).

Compounds exist as:

☑ molecules, such as water, H_2O.

☑ ionic structures

Pure substances and mixtures

In everyday use, the word 'pure' usually means that nothing has been added to a substance.

In chemistry, a **pure** substance contains only one element or compound, e.g.

- pure hydrogen contains only hydrogen molecules
- pure water contains only water molecules.

Most substances are **mixtures** because they contain different elements and/or compounds. Mixture are impure substances.

The components of a mixture are not chemically joined together.

Air

Air is a mixture of:

☑ elements, such as nitrogen, N_2, oxygen, O_2, and argon, Ar

☑ compounds, such as water, H_2O, and carbon dioxide, CO_2.

Worked example

A student heats a sample of solid hexadecanol in a hot water bath. She measures its temperature at regular intervals until after it melts. The graph shows her results.

Explain how the results show whether the hexadecanol is pure or impure. **(2 marks)**

The hexadecanol is pure because it has a sharp melting point, shown by the horizontal part of the heating curve. A mixture would melt over a range of temperatures instead.

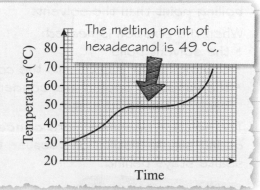

The melting point of hexadecanol is 49 °C.

 Now try this

1 The label on a bottle describes the contents as 'Pure mineral water'.
Explain why the water is not pure in the scientific sense. **(3 marks)**

 2 When it is heated, 18 carat gold melts between 915 °C and 963 °C.
Explain why its melting point is not a single temperature. **(2 marks)**

Distillation

You can separate liquids from mixtures using **distillation**.

 **Simple distillation**

You use **simple distillation** to separate:

☑ a solvent from a solution.

For example, you can separate water from sea water (a mixture of water and dissolved compounds) using this method.

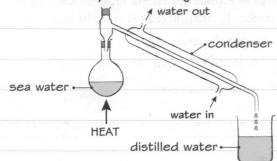

The **condenser** has two tubes, one inside the other:

☑ Cold water runs through the space between the two tubes, keeping the condenser cold.

The cooling water does not mix with the substance being separated.

Practical skills **Fractional distillation**

You use **fractional distillation** to separate:

☑ a liquid from a mixture of **miscible** liquids (liquids that mix completely with each other).

For example, you can separate ethanol from a mixture of ethanol and water using this method.

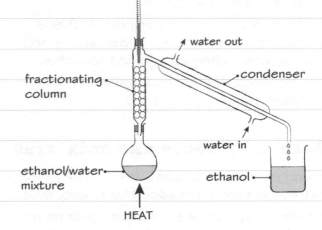

The **fractionating column** used in fractional distillation has a **temperature gradient**:

☑ hottest at the bottom, coldest at the top.

Explaining simple distillation

Simple distillation works because the solute in the solution has a much higher boiling point than the solvent.

When the solution is heated:
- the solvent boils
- solvent vapour passes into the condenser
- the vapour is cooled and condensed back to the liquid state.

The solution becomes more **concentrated** during simple distillation because the solute stays behind.

Explaining fractional distillation

Fractional distillation works because the liquids in the mixture have different boiling points.

When the mixture is heated:
- the mixture boils
- hot vapour rises up the fractionating column
- vapour condenses when it hits the cool surface of the column and drips back
- the fraction with the lowest boiling point reaches the top of the column first
- its vapour passes into the condenser.

If you carry on heating, vapours from fractions with higher boiling points pass to the condenser.

Now try this

1 (a) Explain the function of a condenser in distillation.   **(2 marks)**

(b) State the physical property that allows substances to be separated by distillation.   **(1 mark)**

(c) Describe the temperature gradient in a fractionating column.   **(1 mark)**

Filtration and crystallisation

Filtration and crystallisation are two important techniques for separating mixtures.

 Practical skills ## Filtration

You use **filtration** to separate an insoluble substance from a liquid or a solution. There are two reasons for doing this:

1 to purify a liquid or a solution by removing solid impurities from it, e.g. sand from sea water

2 to separate the solid you want from the liquid it is mixed with, e.g. to separate crystals from a solution after crystallisation.

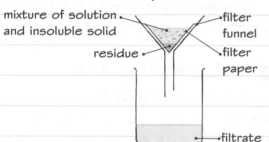

mixture of solution and insoluble solid · · · filter funnel
residue · · · filter paper
· · · filtrate

 Practical skills ## Crystallisation

You use **crystallisation** to produce:
- ✓ solid crystals from a solution

In crystallisation:
- ✓ the solution is heated to remove enough solvent to produce a **saturated** solution (one that cannot hold any more solute)
- ✓ the saturated solution is allowed to cool
- ✓ **crystals** form in the solution
- ✓ the crystals are separated from the liquid and dried.

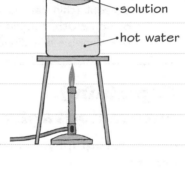
solution
hot water

A hot water bath gives you more control over heating than using a Bunsen burner flame directly on the evaporating basin.

Explaining filtration

Filtration works because the filter paper has tiny pores. These are:
- large enough to let water molecules and dissolved substances through
- small enough to stop insoluble solid particles going through.

Explaining crystallisation

Solubility is the mass of solute that dissolves in a given volume of solvent at a given temperature. Crystallisation works because:
- the solubility of the solute decreases as the saturated solution cools
- crystals form from the excess solute.

Worked example

A student wants to produce some copper sulfate crystals from copper sulfate solution. She heats the solution in an evaporating basin to remove some of the water.

Describe the steps that she should then take to obtain dry crystals of copper sulfate. **(3 marks)**

She should let the solution cool down so that crystals form. She should then decant the remaining liquid and dry the crystals in an oven.

It is safer to make a concentrated or saturated solution than to remove all the water. The crystals will be a better size and shape, and will not spit out of the evaporating basin.

Decanting means pouring the liquid away carefully so that the solid stays behind. The student could use filtration instead to separate the crystals from the liquid. She could pat them dry using filter paper.

Now try this

Glass and dust are insoluble in water but copper chloride is soluble.

1 Outline a method to produce dry copper chloride from a mixture of broken glass, dust and copper chloride powder. **(4 marks)**

Paper chromatography

Paper **chromatography** is used to separate mixtures of soluble substances.

🧪 Practical skills

Paper chromatography

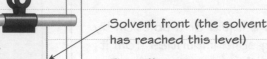

- lid (to stop evaporation of solvent)
- paper
- Drops of the different samples are put onto the paper and allowed to dry. The bottom of the paper is then dipped into a solvent.
- Solvent (this can be water or some other liquid that the samples will dissolve in)
- Solvent front (the solvent has reached this level)
- The different compounds in a sample dissolve to different extents in the solvent.
- More soluble compounds are carried up the paper faster than less soluble ones, so the compounds separate out.

X Y Z

Explaining chromatography

In chromatography, there are two phases:

 1 a **stationary phase** (a substance that does not move)

 2 a **mobile phase** (a substance that moves through the stationary phase – the solvent).

During chromatography:
- each soluble substance in the mixture forms bonds with the two phases
- substances that form stronger attractive forces with the stationary phase stay near the bottom
- substances that form stronger attractive forces with the mobile phase move towards the top.

Using a chromatogram

You can use a **chromatogram** to:
- ✓ distinguish between pure and impure substances (a pure substance will produce only one spot)
- ✓ identify a substance by comparing its pattern of spots with those of a known substance
- ✓ identify substances using R_f values.

🧮 Maths skills — R_f values

$$R_f = \frac{\text{distance travelled by spot}}{\text{distance travelled by solvent}}$$

R_f values have no units. They vary from 0 (spot stays on baseline) to 1 (spot travels with solvent front).

Worked example

What is the R_f value of the lowest spot for sample C in this chromatogram? **(1 mark)**

☐ A 0.20
☒ B 0.40
☐ C 0.44
☐ D 0.90

$R_f = \frac{4 \text{ cm}}{10 \text{ cm}} = 0.4$

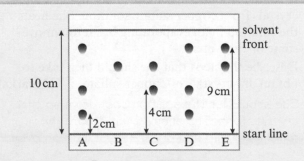

Now try this

1 Calculate the R_f values for the lowest spot in A and the highest spot in E (see above). **(2 marks)**

2 Explain, in terms of attractive forces, why a spot travels far up the paper. **(2 marks)**

Investigating inks

Practical skills You can use **simple distillation** and **paper chromatography** to investigate the composition of inks.

Core practical

You can revise paper chromatography on page 30.

Paper chromatography

Ink is a mixture of coloured substances dissolved in a solvent.

Aims

To investigate the composition of inks using paper chromatography.

Eye protection is important because you should treat all substances in the lab (apart from water straight from the tap) as potentially harmful.

Apparatus

- eye protection
- chromatography paper
- boiling tube with bung
- dropping pipette
- pencil and ruler
- solvent for mobile phase.

You could use a narrow glass **capillary tube** instead of a pipette to obtain tiny drops of ink.

The solvent used depends on the type of ink. For ballpoint pen ink, you may need to use propanone rather than water.

Method

Draw a pencil line near the bottom of the chromatography paper. Apply a small spot of ink, and then place the paper into a boiling tube containing a little solvent. Replace the bung and allow the solvent to travel through the paper.

If you write a **risk assessment**, remember to include hazardous substances and procedures. For example, propanone is flammable and it has a harmful vapour, so you would need to keep it away from Bunsen burner flames and make sure that the lab is well ventilated.

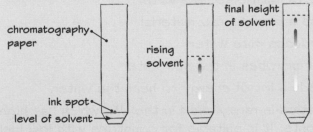

chromatography paper
rising solvent
final height of solvent
ink spot
level of solvent

Simple distillation

You can use simple distillation to separate the solvent in the ink from the coloured substances. The method you use will depend on the apparatus available to you in school.

Revise simple distillation on page 28.

Results

Measure:
- the distance from the pencil line to the solvent front
- the distance from the pencil line to the centre of each spot.

Record the colour of each spot and its R_f value.

Maths skills You could use a suitable table to record the results. Remember:

$$R_f = \frac{\text{distance travelled by spot}}{\text{distance travelled by solvent}}$$

Measure distances in millimetres rather than centimetres.

Now try this

A student uses simple distillation to purify the solvent in a sample of ink. The method that he uses is as follows:
- Add some ink to a boiling tube, then fit a delivery tube and bung.
- Hold the boiling tube with a test-tube holder.
- Heat the ink using a roaring Bunsen burner flame until it is dry.
- Collect the vapour in a test tube.

State **two** ways of improving his method using the same essential equipment.   **(2 marks)**

You could suggest different ways of working. You could also suggest additional apparatus to improve the safety or efficiency of the practical activity.

Drinking water

Waste water and ground water must be treated to make the water **potable** or safe to drink.

Drinking water

Potable drinking water must have:
- low levels of contaminating substances
- low levels of microbes.

Fresh water can come from reservoirs, lakes and rivers. It is likely to contain:
- objects such as leaves and twigs
- insoluble solids such as particles of soil
- microbes, which may cause disease.

The water must be treated to remove these.

Safe but not pure

Tap water is potable but it is not a pure substance. It contains:
- ☑ dissolved salts
- ☑ dissolved chlorine.

Stages in water treatment

The main stages are as follows:

1 **sedimentation** – large insoluble particles sink to the bottom of a tank

2 **filtration** – small insoluble particles are removed by filtering through beds of sand

3 **chlorination** – chlorine gas is bubbled from the water to kill microbes.

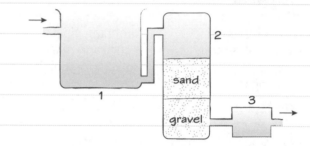

Sea water and distillation

Sea water contains dissolved salts. Their concentration is too high to drink it safely. Sea water can be made potable by simple distillation:
- Filtered sea water is boiled.
- The water vapour is cooled and condensed to form **distilled water**.

Distilled water does not contain dissolved salts, but it contains dissolved gases from the air.

Evaluating distillation

Simple distillation of sea water:
- 👍 uses a plentiful raw material
- 👍 produces pure water
- 👍 kills microbes in the sea water
- 👎 needs a lot of energy to heat the water.

Distillation is rarely used in the UK, which has high fuel costs. It is suitable for countries with low fuel costs, little fresh water, or plentiful sunshine to evaporate the water without using a fuel.

Worked example

Explain why water used for a chemical analysis must not contain any dissolved salts.   **(3 marks)**

Dissolved salts could react with the substances used in the analysis. A product formed in the reaction could interfere with the analysis, giving a false result. If the water used does not contain any dissolved salts, this will not happen.

Many substances used for chemical analysis are used in aqueous solution (dissolved in water).

Dissolved salts may react with a sample, forming insoluble precipitates (see pages 95 and 96). Analytical instruments may detect the dissolved salts, giving a false result. Revise instrumental methods of analysis on page 97.

Now try this

1 Describe what is meant by 'potable'.   **(1 mark)**

2 In the UK, fresh water is piped from reservoirs so that it can be treated to make it safe to drink.
Name and describe the three main stages in water treatment.   **(3 marks)**

Extended response – Separating mixtures

There will be one or more 6-mark question on your exam paper. For these questions, you will need to think scientifically and structure your answer logically, showing how the points you make relate to each other.

You can revise the topics for this question, which is about **states of matter** and **separation techniques**, on pages 26–30.

Worked example

Nitrogen and oxygen are useful substances in air. They can be separated from air by:
- cooling air until water vapour and carbon dioxide solidify
- filtering the chilled air
- further cooling the filtered air to –200 °C
- carrying out fractional distillation.

Substance	Melting point (°C)	Boiling point (°C)
nitrogen	–210	–196
oxygen	–219	–183
water vapour	0	100
carbon dioxide	–78.5	–78.5

The table gives information about four substances found in air. Use the information above, and your knowledge and understanding of separation methods, to explain how nitrogen and oxygen may be separated in this way. **(6 marks)**

If the air is cooled to below –78.5 °C but above –183 °C, the water vapour and carbon dioxide will become solid. The nitrogen and oxygen will stay as gases, so the solid water and carbon dioxide could then be filtered to remove them. This will stop them contaminating the oxygen and nitrogen later in the process.

Use data from the table to suggest the range of temperatures needed to solidify water and carbon dioxide. Use your knowledge and understanding of separation techniques to suggest how they could be separated from oxygen and nitrogen.

When the filtered air is cooled to –200 °C, the oxygen and nitrogen will condense to form a mixture of liquids. This mixture can then be piped into a fractionating column. The column should be warmer at the bottom and colder towards the top.

Use data from the table to describe what happens to the oxygen and nitrogen at –200 °C. Use your knowledge and understanding of fractional distillation to describe the temperature gradient needed in the column.

In the fractionating column, the nitrogen will boil more easily than the oxygen, because it has the lower boiling point. Nitrogen gas can leave from the top of the column and oxygen from the bottom of the column. Depending on the temperature at the bottom, the oxygen could leave as a liquid or a gas.

Use data from the table to explain why oxygen and nitrogen can be separated by fractional distillation. Use your knowledge and understanding to describe where each substance will leave the fractionating column.

You need to show comprehensive knowledge and understanding. Be prepared to apply what you know to a new context. Use the data given to you in a structured way with clear lines of reasoning.

Now try this

Substance	Solubility at 25 °C (g dm⁻³ water)
sodium carbonate	307
calcium carbonate	0.013

Sodium carbonate and calcium carbonate are both white powders at room temperature. The table shows some information about them. A student accidentally makes a mixture of the two powders. Use the information in the table, and your knowledge and understanding of separation methods, to explain how the student could produce separate, dry samples of sodium carbonate and calcium carbonate from this mixture. **(6 marks)**

Acids and alkalis

You can recognise acids and alkalis by their effects on indicator solutions.

Acids

Acids have these properties:

- The pH of their aqueous solutions is less than 7.
- They are a source of **hydrogen ions**, $H^+(aq)$ in solution.

Hydrochloric acid releases H^+ ions:

$$HCl(aq) \rightarrow H^+(aq) + Cl^-(aq)$$

Ethanoic acid releases H^+ ions:

$$CH_3COOH(aq) \rightarrow CH_3COO^-(aq) + H^+(aq)$$

- The higher the concentration of $H^+(aq)$ ions, the lower the pH of the **acidic** solution.

Alkalis

Alkalis have these properties:

- The pH of their aqueous solutions is more than 7.
- They are a source of **hydroxide ions**, $OH^-(aq)$ in solution.

Sodium hydroxide releases OH^- ions:

$$NaOH(aq) \rightarrow Na^+(aq) + OH^-(aq)$$

Ammonia produces OH^- ions in solution:

$$NH_3(g) + H_2O(l) \rightarrow NH_4^+(aq) + OH^-(aq)$$

- The higher the concentration of $OH^-(aq)$ ions, the higher the pH of the **alkaline** solution.

The pH scale

The **pH scale** is a measure of the **acidity** or **alkalinity** of a solution (how acidic or alkaline it is). It goes from 0 to 14. **Neutral** solutions are pH 7.

Indicators are substances that have different colours, depending on their pH. Universal indicator solution or paper is often used to estimate the pH of a solution.

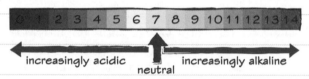

increasingly acidic ← | neutral | → increasingly alkaline

The indicator colour is matched to a colour chart, which shows the pH of each colour.

Worked example

Litmus and phenolphthalein are single indicators. Complete the table to show their colours in acidic and alkaline solutions.   **(4 marks)**

Indicator	In acidic solution	In alkaline solution
litmus	red	blue
phenolphthalein	colourless	pink

Methyl orange is another single indicator (unlike universal indicator, which is a mixed indicator). Single indicators are useful in **titrations**. You can revise titrations on pages 39 and 59.

Methyl orange is:
- red in acidic solutions
- yellow in alkaline solutions.

You will see an orange colour at the end-point of a titration.

Now try this

1 Explain why sodium chloride, NaCl, dissolves in water to produce a solution with a pH of 7, but hydrogen chloride, HCl, dissolves in water to produce a solution with a pH below 7.   **(3 marks)**

2 Potassium reacts with water:

$$2K(s) + 2H_2O(l) \rightarrow 2KOH(aq) + H_2(g)$$

State and explain the effect that the solution formed has on phenolphthalein.   **(2 marks)**

The paper can change colour (as litmus solution does) or stay the same colour.

	Acidic solution	Alkaline solution
Red litmus		
Blue litmus		

3 Copy and complete this table to show the colours of litmus paper in different solutions.   **(2 marks)**

Strong and weak acids

The strength and concentration of an acid determines its pH.

Concentrated versus dilute

For a given volume:
- a **concentrated** solution has a greater amount of dissolved solute particles than a **dilute** solution.

You can change a concentrated solution into a dilute solution by adding more water to it. You can change a dilute solution to a more concentrated solution by:
- dissolving more solute in it
- evaporating some of the water.

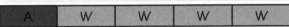 **Maths skills Ratio and proportion**

Diluting a solution and working out its new concentration involves understanding **ratios**.

For example, 1 cm^3 of 2 mol dm^{-3} acid is added to 4 cm^3 of water. This 'bar model' shows the mixture, where A is acid and W is water:

A	W	W	W	W

The ratio of acid to water is 1:4.

The proportion of acid is $\frac{1}{5}$ of the mixture.

The new concentration is:

$$\frac{1}{5} \times 2 = 0.4 \text{ mol dm}^{-3}$$

Strong acids

Hydrochloric acid and sulfuric acid are **strong acids** because they **fully dissociate** into ions in solution. All their molecules release $H^+(aq)$ ions:

$$HCl(aq) \rightarrow H^+(aq) + Cl^-(aq)$$
$$H_2SO_4(aq) \rightarrow 2H^+(aq) + SO_4^{2-}(aq)$$

The concentration of hydrogen ions determines the pH of the solution. The table shows examples:

Concentration of $H^+(aq)$ (mol/dm³)	pH of solution
1	0
0.1	1
0.01	2

- The pH increases by 1 when the H^+ ion concentration decreases by a factor of 10.

You can revise concentration on page 60.

Weak acids

Ethanoic acid is a **weak acid** because it **partially dissociates** into ions in solution. Only a few molecules release $H^+(aq)$ ions:

$$CH_3COOH(aq) \rightleftharpoons CH_3COO^-(aq) + H^+(aq)$$

reaction is **reversible**

At a given concentration of acid:
- a strong acid has a higher concentration of H^+ ions than a weak acid
- a strong acid has a lower pH than a weak acid (see table below).

Acidic solution	pH
0.5 mol dm^{-3} hydrochloric acid	0.3
0.5 mol dm^{-3} ethanoic acid	2.5

You can revise reversible reactions and use of the $\rightleftharpoons$ symbol on pages 67 and 68.

Worked example

0.75 mol dm⁻³ nitric acid has a pH of 0.125. State the pH when 2 cm³ of this acid is mixed with 198 cm³ of water. **(1 mark)**

pH 2.125

The new total volume of acid = (2 + 198)
           = 200 cm³

So the concentration of the acid decreases by (200/2) = 100 times.

If the concentration decreases by two factors of 10 ($10^2 = 100$), the pH increases by 2.

Now try this

1 Explain why a sample of concentrated ethanoic acid may have the same pH as a sample of dilute nitric acid. **(2 marks)**

2 The pH of 0.2 mol dm⁻³ hydrochloric acid is 0.70. Deduce the concentration of hydrochloric acid that has a pH of 2.70. **(1 mark)**

Bases and alkalis

An **alkali** is a soluble **base** – one that will dissolve in water.

Reactions of acids with bases

A **base** is:
- any substance that reacts with an acid to form a salt and water **only**.

Bases are metal oxides and metal hydroxides. In general:

base + acid → salt + water

For example:

$NaOH(aq) + HCl(aq) \rightarrow NaCl(aq) + H_2O(l)$

$CuO(s) + 2HNO_3(aq) \rightarrow Cu(NO_3)_2(aq) + H_2O(l)$

Naming salts

A **salt** forms when hydrogen ions in an acid are replaced by metal ions or ammonium ions. The name of a salt consists of two parts:
- first part – the metal in the base
- second part – from the acid used.

Acid used	Second part
hydrochloric acid, HCl(aq)	chloride
nitric acid, HNO_3(aq)	nitrate
sulfuric acid, H_2SO_4(aq)	sulfate

Reactions of acids with metals

Reactive metals react with acids to produce a salt and hydrogen **only**. In general:

metal + acid → salt + hydrogen

For example:

$Mg(s) + 2HCl(aq) \rightarrow MgCl_2(aq) + H_2(g)$

$2Al(s) + 3H_2SO_4(aq) \rightarrow Al_2(SO_4)_3(aq) + 3H_2(g)$

Reactions of acids with metal carbonates

Metal carbonates react with acids to produce a salt, water **and** carbon dioxide. In general:

metal carbonate + acid → salt + water + carbon dioxide

For example:

$CaCO_3(s) + 2HCl(aq) \rightarrow CaCl_2(aq) + H_2O(l) + CO_2(g)$

Practical skills **Hydrogen**

A **lighted splint** ignites hydrogen with a 'pop'.

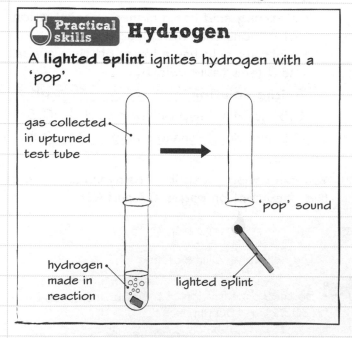

gas collected in upturned test tube

'pop' sound

hydrogen made in reaction

lighted splint

Practical skills **Carbon dioxide**

Carbon dioxide turns **limewater** milky or cloudy white.

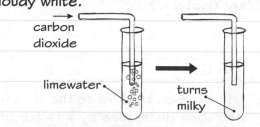

carbon dioxide

limewater

turns milky

Neutralisation

Neutralisation is the reaction between an acid and a base. In an acid–alkali neutralisation, hydrogen ions from the acid react with hydroxide ions from the alkali to form water:

$H^+(aq) + OH^-(aq) \rightleftharpoons H_2O(l)$

Now try this

1 Explain why all alkalis are bases but not all bases are alkalis. **(2 marks)**

2 Give the formulae, and names, for the five salts formed in the reactions shown on this page. **(5 marks)**

Neutralisation

 Core practical

Investigating neutralisation

Aims

To investigate the change in pH when powdered calcium hydroxide, a base, is added to a fixed volume of dilute hydrochloric acid.

 Calcium oxide is corrosive but calcium hydroxide is an irritant, so calcium hydroxide is safer to use.

Apparatus

- eye protection
- beaker
- measuring cylinder
- balance
- spatula
- stirring rod
- white tile
- universal indicator paper and pH colour chart
- dilute hydrochloric acid
- calcium hydroxide powder.

You should choose appropriate apparatus for your practical work. If you are using a measuring cylinder, choose one that is closest to the volume of liquid you need. For example, to measure 50 cm^3, choose a 50 cm^3 measuring cylinder. Do not choose a 25 cm^3 or 100 cm^3 one.

Method

Add some dilute hydrochloric acid to the beaker. Measure and record the pH of the contents of the beaker. Add a small mass of calcium hydroxide powder, stir, and then measure and record the pH again. Repeat until the pH no longer changes.

 You could use a pH meter instead of universal indicator paper. This produces a more precise reading. It will be accurate only if you calibrate it first using standard pH solution.

Results

Use a table to record your results. For example:

Mass of $Ca(OH)_2$ added (g)	pH of mixture
0	
0.3	

 Maths skills Make sure that any units, such as g, go in the column headings only. Do not include units in the main body of the table. Do not forget to take a reading before you add anything!

Analysis

Plot a **titration curve** – a line graph with:
- mass (or volume) added on the horizontal axis
- pH on the vertical axis.

Draw a smooth curve best fit, ignoring **anomalous** results (results that are too high or low).

- ✓ Label each axis with the quantity and unit (if there is one), e.g. 'mass of $Ca(OH)_2$ (g)'.
- ✓ Choose scales so that the plotted points occupy at least 50% of the graph area.
- ✓ Use a sharp pencil for your graph.
- ✓ Plot each point with × (but you might find it more accurate to use + instead).

 **Now try this**

What pH values do you expect at the start, middle and end? Why?

A student adds portions of a powdered soluble base to dilute hydrochloric acid. She estimates the pH each time using universal indicator paper.

 (a) Describe **two** ways she can ensure that her measurements are precise. **(2 marks)**

(b) Describe how she should use the universal indicator paper to estimate the pH. **(3 marks)**

 (c) The student added the base until it was in excess.
Explain the changes in pH during her experiment. **(6 marks)**

Salts from insoluble bases

 Practical skills You can make pure, dry, hydrated copper sulfate crystals, starting from copper oxide.

Core practical

Preparing copper sulfate

Aims

To make a sample of pure, dry, hydrated copper sulfate crystals using copper oxide and sulfuric acid:

$$CuO(s) + H_2SO_4(aq) \rightarrow CuSO_4(aq) + H_2O(l)$$

Apparatus

- eye protection
- beakers
- spatula
- stirring rod
- Bunsen burner
- heat-resistant mat
- tripod and gauze mat
- filter funnel and filter paper
- evaporating basin
- dilute sulfuric acid
- copper oxide powder.

> 'Hydrated' copper sulfate crystals are the familiar blue ones. Their structure includes water molecules as **water of crystallisation**.

> The apparatus is used in the three main stages of this practical:
>
> **Stage 1**: reacting warm sulfuric acid with excess copper oxide powder.
>
> **Stage 2**: filtering the reaction mixture to remove excess (unreacted) copper oxide.
>
> **Stage 3**: producing hydrated copper sulfate crystals by crystallisation with the help of a hot water bath.
>
> You can revise filtration and crystallisation on page 29.

Method

The flow chart describes the main steps needed.

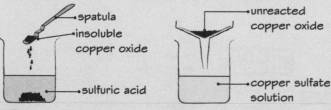

spatula
insoluble copper oxide
sulfuric acid

unreacted copper oxide
copper sulfate solution

copper sulfate crystals formed by evaporating the water

 1 Add excess base to the acid.

2 Filter to remove unreacted copper oxide.

3 Crystallise the copper sulfate solution by heating it or leaving it to stand in a warm place.

> The same method works using a metal but hydrogen, $H_2(g)$, is made rather than water. However, copper does not react with sulfuric acid.

Results

Record your observations at each stage and the appearance of the crystals at the end. This could include the dry mass of the crystals, and their colour, sizes and shapes.

 > Record your observations as you go, rather than trying to remember them afterwards.

Now try this

> **Devise** means that you need to plan or invent a procedure from existing ideas.

Calcium carbonate is an insoluble white solid. It reacts with dilute nitric acid to form calcium nitrate solution:

$$CaCO_3(s) + 2HNO_3(aq) \rightarrow Ca(NO_3)_2(aq) + H_2O(l) + CO_2(g)$$

(a) Describe what you will see when excess calcium carbonate is used in the reaction. **(3 marks)**

(b) Devise a method to prepare a sample of pure, dry, hydrated calcium nitrate crystals. **(6 marks)**

Salts from soluble bases

You should be able to describe how to carry out a **titration** to prepare a pure, dry salt.

Using a soluble base

The flow chart shows the three main steps needed to make a soluble salt from a soluble base.

| Use an acid–base titration to find the exact volume of the soluble base that reacts with the acid. | → | Mix the acid and soluble base in the correct proportions, producing a solution of the salt and water. | → | Warm the salt solution to evaporate the water – this will leave crystals of the salt behind. |

Practical skills — Doing a titration

To carry out a typical titration:

- ✓ Put acid into a **burette**.
- ✓ Use a **pipette** to put a known volume of alkali into a conical flask.
- ✓ Put a few drops of a suitable **indicator** solution, such as phenolphthalein or methyl orange, into the alkali.
- ✓ Record the burette start reading.
- ✓ Add acid to the alkali until the colour changes – the **end-point**.
- ✓ Record the burette end reading.

Maths skills — Mean titres

The **titre** is the volume of acid added to exactly neutralise the alkali:

$$\text{titre} = (\text{end reading}) - (\text{start reading})$$

Concordant titres are identical to each other, or very close together (usually within 0.10 cm³).

You normally calculate the **mean** titre using your concordant results only:

$$\text{mean titre} = \frac{\text{sum of concordant titres}}{\text{number of concordant titres}}$$

You can revise titration calculations on page 61.

Worked example

(a) Explain why titration must be used to produce a salt from dilute hydrochloric acid and potassium hydroxide solution. **(2 marks)**

Titration lets you find the correct proportions of acid and alkali to mix together to produce a solution that contains only a salt and water.

(b) Describe **three** steps needed to obtain an accurate titre during a titration. **(3 marks)**

Swirl the flask continuously during the titration to mix the acid and alkali thoroughly. Near the end-point, add the acid drop by drop, pausing between each addition. Rinse the inside of the flask to make sure all the acid mixes with the alkali.

Obtain accurate readings by clamping the burette vertically and reading it at eye level.

> Potassium hydroxide is a soluble base (an alkali) and the salt (potassium chloride) is also soluble. You can revise solubility rules on page 40.

> When you are making a salt, you find the mean titre, then use the burette to add this volume of acid to the alkali without the indicator.

Apparatus

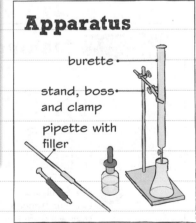

- burette
- stand, boss and clamp
- pipette with filler

Now try this

> The charcoal is still a powder.

1 Instead of carrying out a final titration without the indicator, powdered charcoal can be added at the end-point. This forms strong chemical bonds with the indicator.
Explain how the indicator is then removed from the salt solution. **(2 marks)**

Making insoluble salts

You need to be able to predict, using solubility rules, whether a precipitate will be formed when named solutions are mixed together. If a precipitate is formed, you should be able to name it.

General rules

This is why solutions in practicals are often sodium compounds.

This is very useful to remember.

This follows the first rule at the top.

Soluble	Insoluble
all common sodium, potassium and ammonium salts	
all nitrates	
common chlorides	silver chloride lead chloride
common sulfates	lead sulfate barium sulfate calcium sulfate
sodium hydroxide potassium hydroxide ammonium hydroxide	common hydroxides
sodium carbonate potassium carbonate ammonium carbonate	common carbonates

Revise tests for chloride ions and sulfate ions on page 96.

Worked example

(a) Predict whether a precipitate forms when silver nitrate and sodium chloride solutions are mixed. Name any precipitate formed. **(2 marks)**

a precipitate will form – silver chloride

(b) Name **two** substances that could be used to produce insoluble barium sulfate. **(2 marks)**

barium nitrate and sodium sulfate

Two products form, sodium nitrate and silver chloride, but only silver chloride is insoluble. It forms a cloudy **precipitate** in the mixture:

$AgNO_3(aq) + NaCl(aq) \rightarrow NaNO_3(aq) + AgCl(s)$

All nitrates, and all sodium salts, are soluble. So, to make an insoluble salt **XY**, choose:
- **X** nitrate, and
- sodium **Y**

as your two soluble reactants.

Practical skills — Making insoluble salts

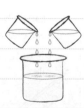  

1 Mix solutions of two substances that will form the insoluble salt.

2 Filter the mixture. The insoluble salt will be trapped in the filter paper.

3 Wash the salt with distilled water.

4 Leave the salt to dry on the filter paper. It could be dried in an oven.

Now try this

1 Predict if a precipitate forms when these solutions are mixed, naming any precipitate formed.
   (a) Sodium carbonate and calcium chloride. **(1 mark)**
   (b) Ammonium nitrate and potassium hydroxide. **(1 mark)**
   (c) Lead nitrate and sodium sulfate. **(1 mark)**

Extended response – Making salts

There will be one or more 6-mark question on your exam paper. For these questions, you will need to think scientifically and structure your answer logically, showing how the points you make are related to each other.

You can revise the topics for this question, which is about **acids, alkalis and salts**, on pages 34–40.

Worked example

Soluble salts can be made by the reactions of acids with insoluble metal compounds. The salt produced depends on the reactants chosen.

Devise a method to prepare pure, dry crystals of zinc sulfate, $ZnSO_4$, from a zinc compound and a suitable acid. Begin your answer by choosing suitable reactants.

You should also write a balanced equation as part of your plan.   **(6 marks)**

I would use zinc carbonate and sulfuric acid. This is the equation for the reaction:

$ZnCO_3 + H_2SO_4 \rightarrow ZnSO_4 + H_2O + CO_2$

I would put some of the acid into a beaker and warm it with a Bunsen burner, tripod and gauze.

I would then use a spatula to add a little zinc carbonate to the warm acid and stir it with a stirring rod. I would repeat this until all the bubbling stopped and some solid was left in the bottom of the beaker.

Next, I would filter the mixture to remove the excess zinc carbonate, using a filter funnel and filter paper. I would collect the zinc sulfate solution in a conical flask.

I would pour the solution into an evaporating basin and leave it on a windowsill for water to evaporate and crystals to form. I would carefully pour away the excess liquid and dry the crystals with filter paper.

Command word: Devise

When you are asked to **devise** something, you need to plan or invent a procedure from existing principles or ideas. These may include familiar chemistry in an unfamiliar context, so make sure that you study carefully any information given to you in the question.

Zinc oxide would also work:

$ZnO + H_2SO_4 \rightarrow ZnSO_4 + H_2O$

Zinc hydroxide would work too:

$Zn(OH)_2 + H_2SO_4 \rightarrow ZnSO_4 + 2H_2O$

You should describe how to make sure that all the acid has reacted. You would not see bubbling if you chose zinc oxide or zinc hydroxide, but you would see unreacted solid left in the beaker when all the acid had reacted.

If you were asked to prepare an insoluble salt by precipitation, you could wash the salt in the filter paper with distilled water at this stage.

In this part of the answer, you could mention heating the solution to evaporate most of the water instead. You would then leave the crystals in a warm place to dry, such as an oven.

You should make sure that your plan is thought out well, with a clear and logical structure. Aim to support it with scientific information and ideas. You could write a method in continuous prose, as here, or you could write it in numbered steps instead. You should apply your knowledge and understanding of the necessary techniques, and use this to answer the question even if you have not made the required salt in school. Include appropriate apparatus without going into irrelevant detail.

Now try this

An insoluble salt can be made by mixing two suitable solutions together to form a precipitate.
Devise a method to prepare a pure, dry sample of insoluble lead(II) chloride, $PbCl_2$.
Begin your answer by choosing suitable solutions and precautions needed for safe working.
You may also write a balanced equation, including state symbols, as part of your plan.   **(6 marks)**

Electrolysis

Electrolysis is used to decompose ionic compounds in the molten state or dissolved in water.

Some key words

An **electrolyte** is:
* an ionic compound in the **molten** state (liquid) or **dissolved** in water.

Electrolysis is:
* a process in which **electrical energy**, from a **direct current** (d.c.) supply, decomposes an electrolyte.

Anions are:
* negatively charged ions that move to the positive **electrode** (anode).

Cations are:
* positively charged ions that move to the negative electrode (cathode).

Moving charges

The ions in an electrolyte are charged particles. An electric current will pass through the electrolyte only if the ions are free to move from place to place.

Remember that ions:

👍 can move about in liquids

👍 can move about in solutions

👎 cannot move about in solids.

You can revise the arrangement and movement of particles in solids and liquids on page 26.

Electrolysis of molten lead bromide

The electrolysis of hot, molten lead bromide produces lead and bromine: $PbBr_2(l) \rightarrow Pb(l) + Br_2(g)$

During electrolysis positive charged ions are attracted to the negative electrode and move to it.

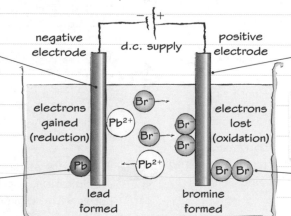

Positively charged ions gain electrons and are **reduced**.

Negatively charged ions are attracted to the positive electrode and move to it during electrolysis.

Remember 'oil rig': oxidation is loss of electrons, and reduction is gain of electrons.

Negatively charged ions lose electrons and are **oxidised**.

Worked example

Molten zinc chloride is electrolysed.

(a) Predict the products formed at each electrode. **(2 marks)**

Zinc forms at the negative electrode and chlorine at the positive electrode.

(b) Write half equations for the reactions occurring at each electrode. **(2 marks)**

negative electrode: $Zn^{2+} + 2e^- \rightarrow Zn$

positive electrode: $2Cl^- \rightarrow Cl_2 + 2e^-$

Zinc chloride (and lead bromide) is a **binary** ionic compound. This means that it consists of two elements only. You must be able to predict the products of electrolysis of such compounds in the molten state.

Electrons are shown as e^- in these equations:
* **Reduction** (gain of electrons) happens at the negative electrode
* **Oxidation** (loss of electrons) happens at the positive electrode.

Now try this

1 (a) Predict the products formed at each electrode during the electrolysis of molten potassium iodide.
    (2 marks)

   (b) Write half-equations for the reactions occurring at each electrode. **(2 marks)**

   (c) Explain at which electrode oxidation occurs. **(2 marks)**

Electrolysing solutions

You need to be able to explain products formed by the electrolysis of compounds in solution.

Ions in a solution

Water is a covalent compound. Some of its molecules naturally form ions:

$$H_2O(l) \rightleftharpoons H^+(aq) + OH^-(aq)$$

You can revise reversible reactions and the use of the $\rightleftharpoons$ symbol on page 67.

The presence of these ions means that a solution of an ionic compound contains:
- cations and anions from the dissolved ionic compound, and
- H^+ and OH^- ions from the water.

Competing ions

During electrolysis of an aqueous solution, all the ions in the electrolyte compete to be discharged and form products.

1 Hydrogen gas is produced if H^+ ions are discharged:

$$2H^+(aq) + 2e^- \rightarrow H_2(g)$$

2 Oxygen gas is produced if OH^- ions are discharged:

$$4OH^-(aq) \rightarrow 2H_2O(l) + O_2(g) + 4e^-$$

At the cathode

metal or hydrogen given off | ions gain electrons

At the negative electrode:
- ✓ hydrogen is produced, **unless**
- ✓ the compound contains ions from a metal less reactive than hydrogen.

In that case:
- ✓ the metal is produced instead.

Copper and silver are below hydrogen in the reactivity series (revise this series on page 46).

At the anode

non-metal (except hydrogen) given off | ions lose electrons

At the positive electrode:
- ✓ oxygen is produced (from OH^- ions) **unless**
- ✓ the compound contains halide ions (Cl^-, Br^- or I^-).

In that case:
- ✓ chlorides produce chlorine, Cl_2
- ✓ bromides produce bromine, Br_2
- ✓ iodides produce iodine, I_2.

Worked example

Predict the products formed at each electrode during the electrolysis of the following concentrated aqueous solutions.

(a) Copper chloride solution. **(1 mark)**

copper at the cathode, chlorine at the anode

(b) Sodium chloride solution. **(1 mark)**

hydrogen at the cathode, chlorine at the anode

(c) Sodium sulfate solution. **(1 mark)**

hydrogen at the cathode, oxygen at the anode

Copper is less reactive than hydrogen, so copper is produced. Chloride ions are present, so chlorine is produced.

Sodium is more reactive than hydrogen, so hydrogen is produced instead of sodium.

Halide ions are not present, so oxygen is produced when OH^- ions are discharged.

Now try this

Which four ions are present, which are discharged and which stay in solution?

1 Dilute sulfuric acid, $H_2SO_4(aq)$, is an electrolyte.
  (a) Give the formulae of all the ions in the electrolyte, and where each comes from. **(3 marks)**

  (b) Predict the product formed at each electrode, and write a balanced half-equation for its formation. **(4 marks)**

 **Practical skills**

Investigating electrolysis

Copper can be purified by the electrolysis of copper sulfate solution using copper electrodes.

Core practical

Electrolysis of copper sulfate

Aims

To investigate the change in mass of the anode and the cathode when copper sulfate solution is electrolysed using copper electrodes.

Apparatus

- eye protection
- beaker
- two strips of copper
- crocodile clips and leads
- d.c. power supply
- copper sulfate solution
- filter paper
- ±0.01 g balance.

Method

Measure and record the mass of each electrode. Connect the apparatus as shown below.

Allow the experiment to run as directed by your teacher, or by your own preliminary experiments. Disconnect the two electrodes, carefully dry them with filter paper, and then weigh them again.

The cathode gains mass as copper ions gain electrons:

$$Cu^{2+}(aq) + 2e^- \rightarrow Cu(s) \quad \text{(reduction)}.$$

The anode loses mass as the copper atoms lose electrons:

$$Cu(s) \rightarrow Cu^{2+}(aq) + 2e^- \quad \text{(oxidation)}.$$

You could bend the strips of copper over the edge of the beaker. Hold them in place using crocodile clips. You can measure the current if you connect an ammeter in series with one of the electrodes.

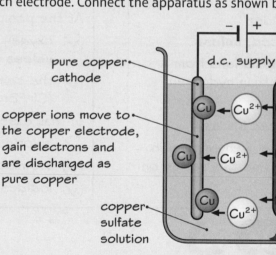

pure copper cathode

copper ions move to the copper electrode, gain electrons and are discharged as pure copper

copper sulfate solution

d.c. supply

impure copper anode

ions are replaced by copper ions from the impure copper anode

impurities form a 'sludge' below the anode

Results

Record the masses of each electrode at the start and end. A table is one way to do this neatly.

Make sure your table has columns to record which electrode is being weighed (at the start and the end) and the mass recorded.

Analysis

Calculate the change in mass of each electrode. Compare these two values and discuss reasons for any difference observed.

The gain in mass of the cathode should equal the loss in mass of the anode.

Now try this

During the electrolysis of copper sulfate solution with copper electrodes, the cathode gains 0.19 g and the anode loses 0.43 g.

(a) The electrodes are rinsed with ethanol, which boils at 78 °C, before drying being dried and weighed. Suggest why they dry more quickly than if water is used. **(1 mark)**

(b) Give **two** reasons why the change in mass of the cathode is less than expected. **(2 marks)**

Extended response – Electrolysis

There will be one or more 6-mark question on your exam paper. For these questions, you will need to think scientifically and structure your answer logically, showing how the points you make are related to each other.

You can revise the topics for this question, which is about **electrolysis**, on pages 42–44.

Worked example

Potassium iodide is an ionic compound. It contains potassium ions, K^+, and iodide ions, I^-. When molten potassium iodide is electrolysed, potassium metal and iodine vapour are formed. Explain how the potassium ions and iodide ions in solid potassium iodide are converted into potassium and iodine by electrolysis.

You may include suitable half-equations in your answer. **(6 marks)**

The potassium iodide must be heated. This is so that its ionic bonds break and it becomes molten. Electrolysis cannot happen in the solid state because the ions are not free to move. It happens in the liquid state because the ions are free to move.

Potassium ions are attracted to the oppositely charged electrode, the negatively charged cathode. At the cathode, the potassium ions gain electrons and become potassium atoms:

$$K^+ + e^- \rightarrow K$$

This is reduction because the ions gain electrons.

Iodide ions are attracted to the positively charged electrode, the anode. Here they lose electrons to form iodine atoms. This is oxidation because the ions lose electrons. Pairs of iodine atoms join together to form an iodine molecule. Overall:

$$2I^- \rightarrow I_2 + 2e^-$$

Command word: Explain

When you are asked to **explain** something, it is not enough just to state or describe it. Your answer **must** contain some reasoning or justification of the points you make. Your explanation **can** include mathematical explanations, if calculations are needed.

In this example Exam skills question, no calculations are needed.

You need to remember what the starting point in the question is. In this case it is solid potassium iodide, not molten potassium iodide. Remember to explain how to melt the compound, and why this is needed.

You should include knowledge and understanding about what happens at the cathode. Include a half-equation and reduction as the gain of electrons.

You should include knowledge and understanding about what happens at the anode. Include a half-equation and oxidation as the loss of electrons. Also mention that a covalent bond forms between two iodine atoms when they share a pair of electrons.

Your answer should show comprehensive knowledge and understanding. Use relevant scientific ideas to support your explanations, including appropriate balanced equations. You should include why potassium iodide must be melted first, then what happens at each electrode. You should leave out information or ideas that are unnecessary to answer the question.

Now try this

Remember to explain whether the lithium and chloride ions will be free to move in the molten potassium chloride. Also explain why this matters.

Lithium-ion batteries are found in many consumer products, including smartphones and electric cars. Lithium is manufactured by the electrolysis of lithium chloride dissolved in molten potassium chloride. During this process, molten lithium rises to the surface and chlorine gas bubbles off.
Explain how the lithium ions and chloride ions in solid lithium chloride are converted into lithium and chlorine by electrolysis. You may include suitable half-equations in your answer. **(6 marks)**

The reactivity series

You can find the relative **reactivity** of a metal by comparing its reactions with other metals.

Reactions with water

Hydrogen is produced if a metal reacts with water:

metal + water → metal hydroxide + hydrogen

For example:

sodium + water → sodium hydroxide + hydrogen

$$2Na(s) + 2H_2O(l) \rightarrow 2NaOH(aq) + H_2(g)$$

In general, the more reactive the metal, the greater the rate of bubbling.

Reactions with acids

Hydrogen is produced if a metal reacts with a dilute acid:

metal + acid → salt + hydrogen

These reactions, and the lab test for hydrogen, are explained in more detail on page 36.

- The rate of reaction is greater in warm acid than in cold acid.
- In general, the more reactive the metal, the greater the rate of bubbling.

Some exceptions

Some reactive metals react unexpectedly slowly with water or acids.

1 Aluminium has a layer of aluminium oxide that stops water reaching the metal below.

2 In the reaction of calcium with dilute sulfuric acid, a layer of insoluble calcium sulfate forms, slowing the reaction.

3 In the reaction of magnesium with water, a layer of sparingly soluble magnesium hydroxide forms, slowing the reaction.

Magnesium reacts vigorously with steam:

$$Mg(s) + H_2O(g) \rightarrow MgO(s) + H_2(g)$$

Metals less reactive than hydrogen do not react with water or dilute acids.

A reactivity series

A **reactivity series** shows elements arranged in order of their reactivity.

	Reaction with	
	Water	Dilute acid
potassium	reacts quickly with cold water	violent reaction
sodium		
calcium		
magnesium	very slow	reaction becoming less vigorous
aluminium	none	
zinc	reacts with steam	
iron		
hydrogen		
copper	no reaction with water or steam	no reaction with dilute acids
silver		
gold		

Remember that hydrogen is not a metal.

Worked example

The diagram shows how five metals react with cold dilute hydrochloric acid and with cold water. Deduce the order of reactivity, starting with the least reactive. **(2 marks)**

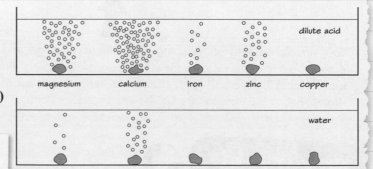

copper, iron, zinc, magnesium, calcium

The rate of bubbling from both diagrams is used to list the metals in the order asked for.

Now try this

1 Using the information in the Worked example:
 (a) Name **one** metal that reacts with cold water. **(1 mark)**

 (b) Name **two** metals that react with dilute hydrochloric acid but do not appear to react with cold water. **(2 marks)**

Metal displacement reactions

You can work out the relative reactivity of metals using displacement reactions.

Displacement

A more reactive metal will **displace** a less reactive metal from its salts in solution. Magnesium is more reactive than copper. It can displace copper from copper sulfate solution:

$$Mg(s) + CuSO_4(aq) \rightarrow MgSO_4(aq) + Cu(s)$$

In this reaction, you observe:

1 the colour of the solution fading as blue copper sulfate is replaced by colourless magnesium sulfate

2 an orange–brown coating of copper forming on the surface of the magnesium

3 an increase in temperature because the reaction is **exothermic** (see page 81).

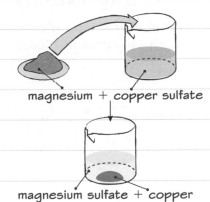
magnesium + copper sulfate

magnesium sulfate + copper

🧪 Practical skills Method for deducing a reactivity series

You can work out a reactivity series for metals by observing what happens when you mix different combinations of metals and their salt solutions. To do this:

☑ add a powdered metal to a test tube or beaker of a metal salt solution (as above right), or

☑ dip a small piece of metal into a metal salt solution on a spotting tile.

Then look for evidence of a reaction, such as a change in colour or temperature.

Some example results

Here are the results of an experiment with three metals and their salt solutions. A ✓ shows where a coating forms on the metal.

	$MgSO_4(aq)$	$ZnSO_4(aq)$	$CuSO_4(aq)$
Mg(s)	✗	✓	✓
Zn(s)	✗	✗	✓
Cu(s)	✗	✗	✗

From the results:

☑ magnesium is the most reactive

☑ copper is the least reactive

☑ a metal does not displace itself.

Worked example

A student adds a spatula of powdered metal to 10 cm³ of copper sulfate solution in boiling tubes. He stirs the mixtures and records the maximum increases in temperature. The table shows his results.

Metal	Temperature increase (°C)	Metal	Temperature increase (°C)
copper	0.0	magnesium	12.5
iron	5.0	zinc	9.5

 (a) Explain, using the results, which metal was the most reactive of the four metals tested. **(2 marks)**

Magnesium was the most reactive because its reaction gave the greatest temperature rise.

(b) Explain the results for copper. **(2 marks)**

There was no temperature change because there was no reaction. A metal does not displace itself from its compounds.

Metals less reactive than copper would also have no reaction with copper sulfate solution.

 ## Now try this

1 Using the table in the Worked example:
 (a) Deduce a reactivity series for the four metals and explain your answer. **(2 marks)**

(b) Silver is less reactive than copper. Explain the expected temperature change if silver is used in the experiment. **(2 marks)**

Explaining metal reactivity

You can explain the reactivity of metals in terms of how easily they form cations.

Forming cations

Metal atoms lose electrons to form **cations** (positively charged ions). For example:

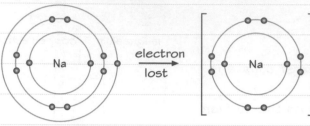

sodium atom, Na   sodium ion, Na$^+$

A full outer shell is obtained when a metal atom loses electrons. The electronic configuration of Na is 2.8.1 but 2.8 for Na$^+$.

OIL RIG

Remember that:

✓ **Oxidation is loss of electrons**

✓ **Reduction is gain of electrons.**

This means that:

✓ atoms are oxidised when they form cations:

$$Na \rightarrow Na^+ + e^-$$

✓ cations are reduced when they form atoms:

$$Na^+ + e^- \rightarrow Na$$

You can revise forming ions on page 10.

Metals in reactions

Metal atoms form cations when metals react with water or dilute acids.

1 Metals react with water to form metal hydroxides, e.g.

$$2Na + 2H_2O \rightarrow 2NaOH + H_2$$

NaOH contains Na$^+$ and OH$^-$ ions.

2 Metals react with dilute acids to form salts, e.g.

$$2Na + 2HCl \rightarrow 2NaCl + H_2$$

NaCl contains Na$^+$ and Cl$^-$ ions.

Metal displacement reactions

Displacement reactions are **redox reactions**:

✓ The atoms of the more reactive metal are **oxidised** – they lose electrons.

✓ The metal cations of the less reactive metal are **reduced** – they gain electrons.

For example, magnesium displaces copper:

$$Mg + CuSO_4 \rightarrow MgSO_4 + Cu$$

You can show this as two **half-equations**:

$$Mg \rightarrow Mg^{2+} + 2e^- \quad \text{oxidation}$$
$$Cu^{2+} + 2e^- \rightarrow Cu \quad \text{reduction}$$

The sulfate ions, SO$_4{}^{2-}$, are unchanged.

Worked example

The list shows four metals arranged in order of decreasing reactivity from left to right.

sodium, aluminium, iron, copper

Which atoms form cations most easily?   **(1 mark)**

☒ **A** Sodium atoms
☐ **B** Aluminium atoms
☐ **C** Iron atoms
☐ **D** Copper atoms

The reactions of metals (with water, dilute acids and salt solutions) show the **relative tendency** of metal atoms to form cations. The easier a metal atom loses electrons, the more reactive the metal is.

Remember that sodium and iron react with water but aluminium, placed between them in the reactivity series, does not. It is protected by a layer of aluminium oxide.

Now try this

1 One way of joining two pieces of railway line together involves the 'thermite reaction'. A mixture of powdered aluminium and iron(III) oxide, Fe$_2$O$_3$, is heated. Aluminium oxide, Al$_2$O$_3$, and molten iron are produced. Use the description of the thermite reaction to help you to explain the difference in reactivity of aluminium and iron.   **(3 marks)**

Metal ores

Most metals are extracted by the reduction of ores found in the Earth's crust.

Ores

Rocks contain metals or their compounds. An **ore** is a rock that contains enough of a metal to make its extraction economical.

Rocks may contain too little metal to make extraction worthwhile (if the cost of extracting the metal is greater than the value of the metal itself). Over time, metal prices may rise and these **low-grade ores** may become useful.

Unreactive metals

Unreactive metals such as gold are placed at the bottom of the reactivity series. They are found in the Earth's crust. They are in their 'native state' uncombined with other elements:

✓ they are not found as **compounds**

but

✓ they may occur naturally as **alloys** (mixtures of metals).

You can revise alloys on page 57.

Oxidation

Oxidation is:

1 the gain of oxygen by a substance, e.g. magnesium is **oxidised** to magnesium oxide in air:

$$2Mg + O_2 \rightarrow 2MgO$$

2 the loss of electrons by a substance, e.g.

$$Mg \rightarrow Mg^{2+} + 2e^-$$

Reduction

Reduction is:

1 the loss of oxygen by a substance, e.g. zinc oxide is **reduced** to zinc when it is heated with carbon:

$$ZnO + C \rightarrow Zn + CO$$

2 the gain of electrons by a substance, e.g.

$$Zn^{2+} + 2e^- \rightarrow Zn$$

Worked example

Copper can be produced from copper oxide, CuO, by heating it with hydrogen gas.

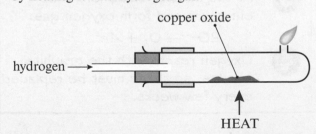

Explain, with the help of a balanced equation, why this is a redox reaction. **(3 marks)**

Copper oxide loses oxygen so it is reduced. At the same time, hydrogen gains oxygen so it is oxidised: $CuO + H_2 \rightarrow Cu + H_2O$

Resistance to corrosion

Metals **corrode** when they react with substances around them, such as air and water. How easily a metal corrodes depends upon how reactive it is:

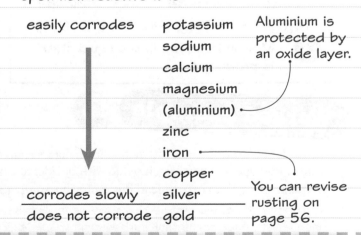

Now try this

1 Explain which metal is more resistant to corrosion – calcium or copper. **(2 marks)**

2 Explain, in terms of loss or gain of oxygen, what happens during oxidation and reduction. **(2 marks)**

3 Explain why platinum is found in its native state in the Earth's crust. **(2 marks)**

Iron and aluminium

Most metals are extracted by the reduction of ores found in the Earth's crust.

Metal extraction

The method used to extract a metal from its ore is related to:
- the **cost** of the extraction process
- the metal's position in the **reactivity series** (shown on the right).

In principle, all metals can be extracted from their compounds using electrolysis. But:
- electricity is needed, which is expensive
- reduction by heating with carbon can be used if a metal is less reactive than carbon
- chemical reactions may be needed to separate silver and gold from other metals.

most reactive	potassium	
	sodium	electrolysis of a molten compound
	calcium	
	magnesium	
	aluminium	
carbon →	zinc	reduction of its oxide using carbon
	iron	
hydrogen →	copper	
	silver	found as the element
least reactive	gold	

Hydrogen can reduce copper oxide to copper, but this is hazardous.

Iron extraction

Iron is less reactive than carbon, so it is produced by reducing iron oxide using carbon. This happens in industrial equipment called a **blast furnace**.
You do not need to know any details about the furnace itself, but at the high temperatures inside it:

1 Iron oxide is reduced by carbon:

iron oxide + carbon → iron + carbon monoxide

$$Fe_2O_3(s) + 3C(s) \rightarrow 2Fe(l) + 3CO(g)$$

2 It is also reduced by carbon monoxide:

iron oxide + carbon monoxide → iron + carbon dioxide

$$Fe_2O_3(s) + 3CO(g) \rightarrow 2Fe(l) + 3CO_2(g)$$

3 Molten iron (iron in the liquid state) is produced.

Aluminium extraction

Aluminium is more reactive than carbon. It is produced by reducing aluminium oxide in an **electrolytic cell**.

1 Aluminium oxide is dissolved in molten **cryolite**. This reduces the temperature needed for electrolysis to happen, when an electric current passes through the mixture.

2 At the **cathode**, aluminium ions gain electrons and are reduced to aluminium atoms:

$$Al^{3+} + 3e^- \rightarrow Al$$

3 At the **anode**, oxide ions lose electrons and form oxygen gas:

$$2O^{2-} \rightarrow O_2 + 4e^-$$

4 Oxygen reacts with the graphite anodes, so these must be replaced every few weeks.

Worked example

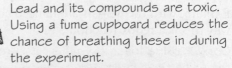

Lead and its compounds are toxic. Using a fume cupboard reduces the chance of breathing these in during the experiment.

A student tries to extract the metals from magnesium oxide and lead oxide. He heats each oxide with powdered carbon. Explain why he can extract lead but not magnesium. **(3 marks)**

Carbon is above lead in the reactivity series but below magnesium. This means that carbon can reduce lead oxide but not magnesium oxide.

It is not enough to write that magnesium is more reactive than lead. You need to include the role of carbon as a **reducing agent**.

Now try this

1 Suggest why tin oxide, but not calcium oxide, can be reduced by carbon. **(3 marks)**
2 Explain why electrolysis is not used to extract iron from iron oxide. **(2 marks)**

Biological metal extraction

There are biological methods for extracting metals, including use of plants or bacteria.

Using scrap iron

Copper is a valuable metal but we are running out of **high-grade** copper ores. It is expensive to extract copper from **low-grade** ores using traditional methods. Therefore, other methods are being researched.

Scrap iron can be used to produce copper from solutions of copper salts:

$$\text{iron} + \text{copper sulfate} \rightarrow \text{iron(II) sulfate} + \text{copper}$$

$$Fe(s) + CuSO_4(aq) \rightarrow FeSO_4(aq) + Cu(s)$$

Iron is more reactive than copper. It can displace copper from copper compounds. You can revise metal displacement on page 47.

Grades of ore

Grade of ore	Proportion of metal or metal compound
high	high
↓	↓
low	low

Compared with high-grade ores, low-grade ores:

👍 are more common because most high-grade ores have already been used
👎 are less profitable
👎 use more energy
👎 produce more waste when used.

Phytoextraction

Phytoextraction is a biological method of metal extraction that uses plants.

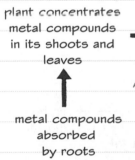 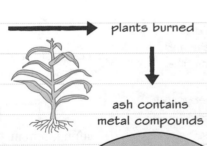

plant concentrates metal compounds in its shoots and leaves

plants burned

metal compounds absorbed by roots

ash contains metal compounds

Bioleaching

Bioleaching is a biological method of metal extraction that uses bacteria. Copper can be extracted from copper sulfide, CuS, in the following way:

✓ the bacteria oxidise sulfide ions, S^{2-}
✓ copper sulfide ores break down
✓ Cu^{2+} ions are released.

The solution that the bacteria produce is called a **leachate**. This leachate contains:

✓ a high concentration of metal ions.

Scrap iron can be used to obtain copper from the leachate.

Worked example

Copper can be produced from ores by two methods:
Method A involves mining the ore, purifying it and heating the purified ore with carbon.
Method B is phytoextraction using plants.
Evaluate these two extraction methods. **(6 marks)**

When you are asked to evaluate something, make sure that you give its advantages *and* disadvantages. Include a conclusion.

Method A is quicker but supplies of high-grade ores are limited. It uses more energy and produces large amounts of waste rock.

Method B produces ash with a high concentration of copper from low-grade ore. It does not cause pollution as a result of mining. On the other hand, it takes a lot longer because the plants have to grow and it uses more land.

Overall I think that method B is better because it conserves supplies of high-grade ores.

Now try this

1 Describe how phytoextraction works. **(4 marks)**
2 Describe **one** advantage and **one** disadvantage of extracting metals using bacteria. **(2 marks)**

Recycling metals

Recycling metals rather than extracting them from ores has economic implications. It can also preserve the environment and the supply of limited raw materials.

Extracting metals

Extracting metals from their ores:

👎 uses up limited resources

👎 uses a lot of energy

👎 damages the environment.

Recycling reduces these disadvantages. Used metal items are collected. Rather than throwing them away, these are taken apart. The metal is melted down to make new items.

 VS ## Recycling

Recycling metals means:

👍 metal ores will last longer

👍 less energy is needed

👍 fewer quarries and mines are needed

👍 less noise and dust are produced

👍 less land is needed.

Worked example

The flow chart shows the main stages in extracting aluminium from its ore.
Use it to suggest the benefits of recycling aluminium. **(3 marks)**

Less waste rock will be produced from mining.

Aluminium oxide will not need separating and purifying from aluminium ore, which will save energy. Less carbon dioxide will be emitted because less fuel will be needed for heat and electricity and because carbon dioxide is produced from the electrolysis.

You need to be specific in your answer. To write that recycling is 'better for the environment' does not give enough detail.

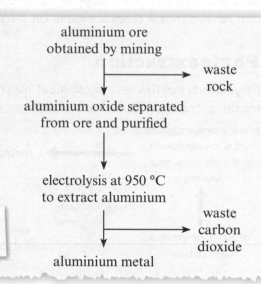

aluminium ore
obtained by mining

→ waste rock

aluminium oxide separated
from ore and purified

electrolysis at 950 °C
to extract aluminium

→ waste carbon dioxide

aluminium metal

Drawbacks of recycling

👎 Used metal items must be collected and transported to the recycling centre.

👎 Different metals must be removed from used items and sorted.

👎 Recycling saves different amounts of energy, depending on the metal involved.

Recycling metals saves energy

Different amounts of energy are saved by recycling metals compared with extracting them from ores:

Metal	Percentage energy saved
aluminium	94
copper	86
iron and steel	70

Now try this

1 Describe **two** ways in which recycling copper, rather than extracting it from copper ores, can reduce pollution. **(2 marks)**

2 Suggest **two** reasons why more energy is saved when aluminium is recycled than when steel is recycled. **(2 marks)**

3 Explain why recycling metals conserves the supplies of metal ores. **(2 marks)**

Life-cycle assessments

You need to be able to describe the basic principles involved in carrying out a life-cycle assessment.

Cradle to grave

A **life-cycle assessment** (LCA) of a product is a 'cradle-to-grave' analysis of its impact on the environment. It includes these stages:

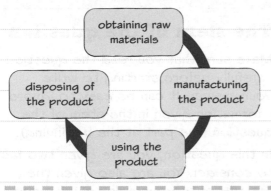

obtaining raw materials

manufacturing the product

using the product

disposing of the product

Data for an LCA

An LCA is likely to need data on these factors at most or all stages:

- ✓ the use of energy
- ✓ the release of waste materials
- ✓ transport and storage.

An LCA is also likely to need data on:

- ✓ whether the raw materials needed are renewable or non-renewable
- ✓ whether any of the product can be recycled or re-used
- ✓ how the product is disposed of.

An example of a life cycle

There are many detailed stages in the life cycle of a product, e.g. for a car.

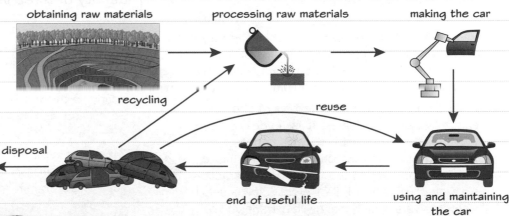

obtaining raw materials

processing raw materials

making the car

recycling

reuse

disposal

end of useful life

using and maintaining the car

Worked example

The table shows information about the energy needed in the life cycle of a polyester shirt.

Stage	Percentage of lifetime energy used
obtaining raw materials	5
manufacturing the shirt	14
using the shirt	80
disposing of the shirt	1

Discuss the use of energy in the life cycle of the shirt, and suggest how it may be improved. **(4 marks)**

Almost one-fifth of the energy used is to do with making the shirt. Very little is to do with disposing of it at the end of its life. Most of the energy used is to do with using the shirt, probably due to washing and ironing it. Energy use could be reduced by washing at low temperatures and drying it outside.

The question shows only part of the data that should be available to make a life-cycle assessment, e.g. no information is given about the environmental impact of each stage, or the lifespan of the shirt.

Now try this

1 State the four main stages in a life-cycle assessment. **(1 mark)**

2 Life-cycle assessments may identify alternative materials that can be used with less impact. Give one other reason for carrying out a life-cycle assessment. **(1 mark)**

Extended response – Reactivity of metals

There will be one or more 6-mark question on your exam paper. For these questions, you will need to think scientifically and structure your answer logically, showing how the points you make are related to each other.

You can revise the topics for this question, which is about **metal reactivity** and **metal extraction**, on pages 46–50.

Worked example

Some metals are found as uncombined elements, but most metals are extracted from ores found in the Earth's crust. Different metals are extracted using different methods.

Explain how the method used to obtain a metal is related to its position in the reactivity series and to the cost of the extraction process. In your answer, refer to aluminium, iron and gold as examples.

(6 marks)

Gold is placed at the bottom of the reactivity series because it is an unreactive metal. It is found in the Earth's crust uncombined with other elements so the cost of extracting gold is low.

Aluminium is placed near the top of the reactivity series, because it is a very reactive metal. It is found combined with other elements, such as oxygen. Aluminium is more reactive than carbon, so it must be extracted using electrolysis. A lot of electricity is used in electrolysis, which makes it expensive to extract aluminium from its ore.

Iron is placed in the middle of the reactivity series, because it is less reactive than aluminium but more reactive than gold. Iron is also less reactive than carbon. Iron can be extracted by heating its ore with carbon. Although electrolysis could be used instead, carbon is cheaper than electricity.

The stem of the question

Make sure that you read the question carefully before starting to write your answer. It can be easy to ignore information given in the stem of the question (the part at the beginning).

In this question, you are given two factors to consider. You are also given the identity of three metals to write about.

You could begin your answer with any of the metals. The extraction of gold is simple to explain so it is covered first here. In your answer, refer clearly to the position of gold in the reactivity series, and also to the cost of extracting it.

You should include knowledge and understanding about why aluminium must be extracted using electrolysis. You should also include why this is expensive. You need to explain why heating with carbon is unsuitable. You can do this by referring to the position of carbon in the reactivity series.

You might point out that electrolysis could still be used to extract iron. You can then explain that heating with carbon is cheaper.

You should think clearly about how to present your ideas before you start. Aim for a clear and logical structure. You can support this with scientific information and ideas. You should leave out information or ideas that are unnecessary to answer the question.

Now try this

You are provided with these metal powders and solutions:

Metal powder	Solution
copper	copper sulfate
magnesium	magnesium sulfate
zinc	zinc sulfate

Explain how you could use these substances in displacement reactions to find the order of reactivity of the three metals. In your answer, include the expected results and how you would use them to show the correct order of reactivity. You should use equations as part of your answer. **(6 marks)**

Transition metals

Most metals are **transition metals**, placed between groups 2 and 3 in the periodic table.

Physical properties of transition metals

Typical properties include:

- high melting point (except mercury, which is in the liquid state at room temperature)
- high density (they have a high mass for their volume).

Transition metals are stronger and harder than the metals in groups 1 and 2, so they are often more suitable as construction materials.

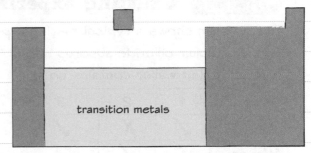

transition metals

Chemical properties of transition metals

Typical properties include:

- formation of coloured compounds
- catalytic activity.

Transition metals and their compounds are often useful as **catalysts**. These substances increase the rate of a chemical reaction **without**:

- altering the products of the reaction
- changing chemically
- changing in mass at the end of the reaction.

You can revise catalysts on page 78.

Group 1 and 2 properties

Typical properties include:

- ✓ relatively low melting point
- ✓ relatively low density
- ✓ formation of white or colourless compounds
- ✓ lack of catalytic activity.

These properties are the opposite of those for transition metals.

You can revise the properties of group 1 elements in more detail on page 72.

Worked example

Iron reacts with hydrochloric acid to produce hydrogen and green iron(II) chloride solution.

(a) Write a balanced equation for the reaction. **(2 marks)**

$Fe(s) + 2HCl(aq) \rightarrow FeCl_2(aq) + H_2(g)$

(b) Explain how you can tell that iron is a transition metal. **(2 marks)**

Transition metals form coloured compounds and iron(II) chloride is described as green.

(c) Give one example of the use of iron as a catalyst. **(1 mark)**

Iron is the catalyst in the Haber process.

Chloride ions have the formula Cl⁻ so the charge on the iron ion in $FeCl_2$ must be 2+.

The Roman number 2 in brackets shows that the ion is Fe^{2+} – not Fe^{3+}, which is the iron(III) ion.

Note that iron(III) chloride, $FeCl_3$, forms a different coloured solution (orange–brown). Iron also has:

- a high melting point (1538 °C)
- a high density (7870 kg/m³).

The Haber process is a reversible reaction between nitrogen and hydrogen.

You can revise this industrial process on page 67.

Now try this

1 Give four typical properties of transition metals. **(4 marks)**

2 A student has a white powder and a red powder. Explain how he can decide which one is copper(I) oxide and which one is aluminium oxide. **(3 marks)**

Rusting

The oxidation of metals results in **corrosion**. The corrosion of iron and steel is called **rusting**.

 **A rusting experiment**

The diagram shows a typical rusting experiment.

Tube 1: calcium chloride absorbs water.

Tube 2: boiled water contains no air.

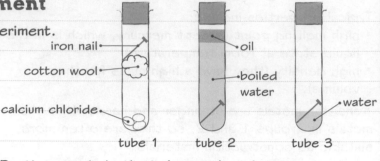

Tube	1	2	3
Air	✓	✗	✓
Water	✗	✓	✓
Rusted?	✗	✗	✓

Rusting needs both air (oxygen) **and** water:

iron + oxygen + water → hydrated iron(III) oxide

Rust prevention

Rust prevention may rely on keeping oxygen **and** water away from the surface of the iron or steel. Ways to do this include:
- painting
- using oil or grease
- coating with plastic
- coating with another metal.

Rust prevention may keep oxygen away:
- Store the item in a vacuum container.

Rust prevention may keep water away:
- Put the item in a container with a **desiccant** (which absorbs water vapour).

The method you choose depends on factors such as cost and suitability. The inside of a food can is protected by a thin layer of tin – you would not want to use oil or grease.

Electroplating

A metal object can be **electroplated** to:

✓ improve its appearance

✓ improve its resistance to corrosion.

A thin layer of an unreactive metal such as nickel, silver or gold is deposited on the surface of the metal object, keeping air and water out.

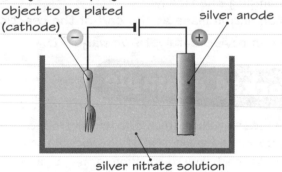

Worked example

Iron and steel objects can be coated with a thin layer of zinc to prevent them from rusting.

(a) Give the name of this process.　**(1 mark)**

galvanising

(b) Explain how this method of rust prevention works.　**(3 marks)**

Zinc is more reactive than iron, so it corrodes before the iron does. The layer of zinc also stops air and water reaching the iron beneath.

Metal objects are galvanised by dipping them in molten zinc or electroplating with zinc. Car body parts are galvanised, then painted for extra protection and to make them attractive.

Galvanising works even if the zinc layer is scratched because it is sacrificial protection.

Other metals can be used for sacrificial protection. The metal chosen must:
- be more reactive than iron, but
- react slowly (not rapidly) with water.

Magnesium is another suitable sacrificial metal.

Now try this

1　Explain why rusting is an example of an oxidation reaction.　**(2 marks)**
2　Describe how painting prevents rust.　**(2 marks)**

Alloys

An **alloy** is a mixture of a metal with one or more other elements. These other elements are usually other metals or carbon.

Steel

Pure iron is too soft for most purposes. It is mixed with carbon, and other metals, to produce alloys called **steels**.

There are many types of steel with different uses depending on their properties. The most common steels are **carbon steels**. They:

- consist of iron with up to 2.0% carbon.
- are harder and stronger than iron alone.

Stainless steel

Stainless steel is an iron alloy that resists corrosion. It is used to make cutlery, washing machine drums and dishwashers.

Stainless steel contains chromium:

☑ an invisible thin layer of chromium oxide forms on the surface of the steel

☑ **it** stops air and water reaching the iron.

The layer quickly reforms if it is scratched.

Some alloys and their uses

Alloy	Main metal	Mixed with	Typical properties	Uses
carbon steel	iron	carbon	hard, strong	buildings, bridges, cars
magnalium	aluminium	magnesium	low density	car and aeroplane parts
jewellery gold	gold	copper	attractive, resistant to corrosion (so it stays shiny)	jewellery
brass	copper	zinc	hard, resistant to corrosion, good electrical conductor	electrical plugs, coins

Converting pure metals into alloys often increases the strength of the metal.

For example, pure '24-carat' gold is too soft for most purposes, so it is mixed with copper:

Carats	24	22	18	14
Percentage of gold	100	92	75	58
Percentage of copper	0	8	25	42
Relative hardness	0.31	0.48	0.72	1.0

softest ⟶ hardest

Worked example

Brass is an alloy of copper and zinc. The diagrams show examples of how the atoms are arranged in pure copper and in brass.

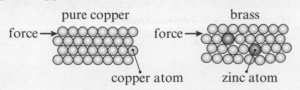

pure copper   brass

force →   force →

copper atom   zinc atom

Explain why brass is stronger than copper.

(2 marks)

In pure copper, layers of atoms can slide over each other relatively easily. However, the larger zinc atoms disrupt the layers in brass. This means that the atoms cannot slide over each other so easily.

Now try this

Think about the physical and chemical properties of copper and gold.

1 Suggest **two** properties, unrelated to hardness, that explain why copper is a suitable metal to mix with gold for use in jewellery.
(2 marks)

Extended response – Alloys and corrosion

There will be one or more 6-mark question on your exam paper. For these questions, you will need to think scientifically and structure your answer logically, showing how the points you make are related to each other.

You can revise the topics for this question, which is about **transition metals**, **alloys** and **corrosion**, on pages 55–57.

Worked example

A student is on holiday at the seaside. He sees that steel railings by the beach are often in contact with seawater, and are very rusty.

When the student returns to the hotel in the town, he sees that the steel railings there are much less rusty.

The student predicts that seawater makes steel rust faster than rainwater does.

Devise an experiment that the student could do to test his prediction. **(6 marks)**

Instead of pieces of steel railings, the student could use steel nails. He could use three nails and put each one into a test tube.

He could put one nail in seawater and one in rainwater. Both nails should only be half covered with water because rusting needs both oxygen and water to happen. The third nail could be left without any water to act as a control.

The student could then leave all three nails in their tubes for the same amount of time, such as 1 week. After this time, the student could observe and record the appearance of each nail, looking for signs of rusting such as an orange–brown coating or flakes of rust in the test tube.

If the seawater nail were rustier than the rainwater nail, the student's prediction would be correct. To check if his results are repeatable, the student could use more than one set of three test tubes, or do his experiment again.

 Notice that this question does **not** ask you to devise an experiment to find out what **causes** rusting. It assumes that you already know this. The question is asking you to plan a procedure to do with rusting. You should base this on information given to you, and your knowledge and understanding of experiments.

A clearly annotated diagram is also an effective way to describe a procedure.

 You should recognise that it is not sensible to experiment with the actual railings. Instead, you should choose small objects made from steel.

 You could be more precise here, and state that the same volume of water should be used. You could also state clearly that the three nails should be identical.

Make sure that you give a method for measuring the amount of rusting. There are other methods you could use, including measuring and recording the dry mass of the nails (after wiping the surface with filter paper) at the start and end.

 If you get similar results more than once using the same method and equipment, the results are **repeatable**. If someone else gets similar results using a different method and equipment, the results are **reproducible**.

Now try this

An aluminium alloy commonly used to make aircraft contains 4.4% copper, with smaller amounts of other elements. Explain why this alloy is used for aircraft rather than pure aluminium or steel. **(6 marks)**

Metal	Density (kg/m³)	Relative strength
aluminium	2700	1.0
aluminium alloy	2800	4.7
steel	7800	11.8

Think about the properties an aircraft material should have. What properties, in addition to those given, would an aluminium alloy need and why?

Accurate titrations

 Practical skills You can carry out an accurate acid–alkali **titration** to determine the concentration of one of two reacting solutions.

Core practical

An acid–alkali titration

Aims

To carry out an accurate acid–alkali titration using a burette, pipette and a suitable indicator.

Apparatus

- eye protection
- burette
- stand, boss and clamp
- funnel
- white tile
- pipette and pipette filler
- conical flask
- beakers
- 0.1 mol dm⁻³ (approx.) hydrochloric acid
- 0.100 mol dm⁻³ sodium hydroxide solution.

Method

1. Use a beaker and funnel to put dilute hydrochloric acid into the burette.
2. Use the pipette and pipette filler to add 25.0 cm³ of sodium hydroxide solution to the conical flask. Then add a little indicator.
3. Put the flask on a white tile underneath the burette.
4. Record the burette start reading.
5. Add acid to the alkali until the end-point.
6. Record the burette end reading.
7. Calculate the titre (see Results section).
8. Repeat steps 1 to 7 until concordant titres are obtained.

Results

Record the burette readings in a suitable table, such as the one below:

titre = (end reading) – (start reading)

Titration run	1	2	3	4
End reading (cm³)	24.00	46.60	22.70	45.65
Start reading (cm³)	0.75	24.00	0.00	22.70
Titre (cm³)	23.25	22.60	22.70	22.95

Analysis

Tick concordant titres and calculate the mean.

Suitable indicators include phenolphthalein and methyl orange.

Instead of a boss and clamp you may be given a burette holder to use. A volumetric pipette is more accurate and repeatable than using a measuring cylinder.

These solutions and their concentrations are given as examples only. You may be given a different acid or alkali, its concentration may be different, or you may be asked to put alkali in the burette and acid in the flask instead.

To work safely and accurately, you should:
- wear eye protection
- fill the burette no higher than eye level
- make sure that the burette is vertical
- swirl the flask while adding the acid to it
- add acid drop by drop near the end-point (when the indicator changes colour)
- read the burette at eye level
- record readings to the nearest 0.05 cm³.

0 —

← 0.75 cm³

1 —

Usually at least two titres within 0.10 cm³ of each other.

Now try this

Look at page 34 to help you to answer (c).

Refer to the table of results opposite when you answer these questions.

(a) Suggest a reason that explains why the first titration run is usually a rough run, with its titre an anomalous result. **(1 mark)**

(b) Identify the concordant titres, and use them to calculate the mean titre. **(2 marks)**

(c) State the colour change observed at the end-point in this experiment, if phenolphthalein indicator is used. **(1 mark)**

Concentration calculations

You need to be able to calculate the concentrations of solutions in mol dm^{-3}. You also need to know how to convert between concentration in g dm^{-3} and concentration in mol dm^{-3}.

Amount, volume and concentration

You use this equation to calculate the concentration of a solution in mol dm^{-3}:

$$\text{concentration (mol dm}^{-3}) = \frac{\text{amount of solute (mol)}}{\text{volume of solution (dm}^3)}$$

LEARN IT!
IT'S NOT ON THE EQUATIONS LIST

🖩 Maths skills **Units**

The unit for **amount of substance** is the mole, mol. You can revise moles on page 24. The unit mol dm^{-3} means 'moles per cubic decimetre'.

You may also see it written as mol/dm^3.

🖩 Maths skills **Rearranging**

You need to be able to change the subject of an equation. For example:

☑ amount of solute = concentration × volume

☑ volume = $\dfrac{\text{amount of solute}}{\text{concentration}}$

Worked example

26.9 g of copper chloride, $CuCl_2$, is dissolved in 125 cm^3 of water.
Calculate the concentration of the solution in mol dm^{-3}.
(Relative atomic masses: Cu = 63.5, Cl = 35.5)

(3 marks)

M_r of $CuCl_2$ = 63.5 + (2 × 35.5) = 134.5

amount of $CuCl_2$ = $\dfrac{\text{mass}}{M_r}$ = $\dfrac{26.9}{134.5}$ = 0.200 mol

volume = $\dfrac{125}{1000}$ = 0.125 dm^3

concentration = $\dfrac{0.200 \text{ mol}}{0.125 \text{ dm}^3}$ = 1.60 mol dm^{-3}

🖩 Maths skills
The values in the question are given to three significant figures, so give the answer to three significant figures.

🖩 Maths skills
If you are not given the relative atomic masses, A_r, in a question, look for them in a periodic table.

🖩 Maths skills
Remember: 1 dm^3 = 1000 cm^3
So divide by 1000 to go from cm^3 to dm^3.

You can revise these conversions on page 23.

🖩 Maths skills
Check answers to simple calculations using estimates. In the last part here, the top number is only slightly larger than the bottom number, so the answer should be a little more than 1.

Golden rule: conversions

You can easily convert between mol dm^{-3} and g/dm^3 using the M_r of the solute:

☑ mol dm^{-3} → g dm^{-3} – multiply the concentration by the M_r

☑ g dm^{-3} → mol dm^{-3} – divide the concentration by the M_r.

Worked example

Calculate the concentration, in g dm^{-3}, of a 0.25 mol dm^{-3} solution of sodium hydroxide, NaOH (relative formula mass = 40). **(1 mark)**

concentration = 0.25 mol dm^{-3} × 40
= 10 g dm^{-3}

Now try this

Calculate the amount in mol first, then use the M_r to convert amount to mass.

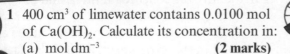

1 400 cm^3 of limewater contains 0.0100 mol of $Ca(OH)_2$. Calculate its concentration in:
(a) mol dm^{-3} **(2 marks)**
(b) g dm^{-3}. **(2 marks)**

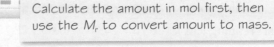

2 Calculate the mass of sodium hydroxide needed to make 250 cm^3 of a 0.100 mol dm^{-3} solution (M_r = 40). **(3 marks)**

Titration calculations

You need to be able to use titration results to calculate concentrations or volumes of solutions.

Worked example

In a titration, 25.00 cm³ of sodium hydroxide solution reacts with 24.00 cm³ of 0.10 mol dm⁻³ hydrochloric acid, HCl.
Calculate the concentration of the sodium hydroxide solution, NaOH, in mol dm⁻³.

$$HCl + NaOH \rightarrow NaCl + H_2O$$ **(3 marks)**

volume of HCl = $\dfrac{24.00 \text{ cm}^3}{1000}$ = 0.024 dm³

number of moles of HCl = concentration × volume

= 0.10 mol dm⁻³ × 0.024 dm³ = 0.0024 mol

From the equation, 1 mol of HCl reacts with 1 mol of NaOH, so there will be 0.0024 mol of NaOH.

volume of NaOH = $\dfrac{25.00 \text{ cm}^3}{1000}$ = 0.025 dm³

concentration of NaOH = $\dfrac{\text{number of moles}}{\text{volume}}$

= $\dfrac{0.0024 \text{ mol}}{0.025 \text{ dm}^3}$ = 0.096 mol dm⁻³

> Notice that you do not need to use any relative atomic masses or relative formula masses in these calculations.

Worked example

25.00 cm³ of 0.80 mol dm⁻³ sodium hydroxide solution, NaOH, reacts with 1.0 mol dm⁻³ sulfuric acid, H₂SO₄.
Calculate the volume of acid needed to neutralise the sodium hydroxide.

$$H_2SO_4 + 2NaOH \rightarrow Na_2SO_4 + 2H_2O$$ **(3 marks)**

> Notice that 1 mol of H₂SO₄ reacts with 2 mol of NaOH. This is different from the situation when hydrochloric acid is used, when 1 mol of HCl reacts with just 1 mol of NaOH:
>
> $$HCl + NaOH \rightarrow NaCl + H_2O$$

volume of NaOH = $\dfrac{25.00 \text{ cm}^3}{1000}$ = 0.025 dm³

number of moles of NaOH = concentration × volume

= 0.80 mol dm⁻³ × 0.025 dm³ = 0.020 mol

> Divide by 1000 to convert from cm³ to dm³.

From the equation, 1 mol of H₂OS₄ reacts with 2 mol of NaOH, so there will be 0.010 mol of H₂SO₄.

> The 'mole ratio' between sulfuric acid and sodium hydroxide solution is 1:2 – this is why the number of moles of sulfuric acid is half the number of moles of sodium hydroxide.

volume of H₂SO₄ = $\dfrac{\text{number of moles}}{\text{concentration}}$

= $\dfrac{0.010 \text{ mol}}{1.0 \text{ mol dm}^{-3}}$ = 0.010 dm³ (or 10 cm³)

> Multiply by 1000 to convert from dm³ to cm³.

Now try this

1 25.00 cm³ of potassium hydroxide solution, KOH, reacts with 28.00 cm³ of 0.200 mol dm⁻³ nitric acid, HNO₃:

$$HNO_3 + KOH \rightarrow KNO_3 + H_2O$$

Calculate the concentration of potassium hydroxide in mol dm⁻³.   **(3 marks)**

2 25 cm³ of 0.20 mol dm⁻³ sodium hydroxide solution reacts with 0.50 mol dm⁻³ hydrochloric acid.
Calculate the volume of acid needed to neutralise the sodium hydroxide.   **(3 marks)**

Percentage yield

The mass of a product made in a chemical process is called the **yield**.

Theoretical yield

In any chemical reaction:
- no atoms are gained or lost
- the total mass stays the same.

This means that, for a given mass of a limiting reactant, you can make only a maximum mass of a product. This is the **theoretical yield**.

You can revise the law of conservation of mass on page 21 and reacting mass calculations using balanced equations on page 22.

Actual yield

The **actual yield** is the mass of product that you really get at the end of a chemical process. This is always less than the theoretical yield.

You can calculate the **percentage yield** using:

$$\text{percentage yield} = \frac{\text{actual yield}}{\text{theoretical yield}} \times 100$$

LEARN IT!
IT'S NOT ON THE EQUATIONS LIST

Yield	0%	100%
Product	none made	none lost

Obtaining less product than expected

incomplete reactions:
- reaction has not finished
- reaction reaches equilibrium

← reasons for not obtaining the theoretical yield →

practical losses during the experiment:
- losses during purification, e.g. filtration
- losses during transfers, e.g. liquid left behind in containers

side reactions:
- competing, unwanted reactions, so by-products are also made

Worked example

Copper carbonate decomposes when it is heated. Copper oxide and carbon dioxide are formed:

$$CuCO_3(s) \rightarrow CuO(s) + CO_2(g)$$

A student carries out four experiments. She varies the mass of copper carbonate used each time. The table shows her results.

Experiment	Theoretical yield (g)	Actual yield (g)
1	0.5	0.4
2	1.0	0.8
3	1.5	1.2
4	2.0	1.5

(a) Suggest one reason that explains why the actual yields are less than the theoretical yields. **(1 mark)**

Some of the copper carbonate may not have decomposed to form copper oxide.

(b) Explain which experiment gave a result that was anomalous. **(2 marks)**

Experiment 4 because the actual yield did not follow the trend. You would expect to get 1.6 g of copper oxide from 2.0 g of copper carbonate. The student got less than this.

(c) Calculate the percentage yield for experiment 3. **(1 mark)**

$$\text{percentage yield} = \frac{\text{actual yield}}{\text{theoretical yield}} \times 100$$

$$= \frac{1.2}{1.5} \times 100 = 80\%$$

Check that your answer makes sense. It cannot be higher than 100%. An answer of 125% shows that you used the equation upside down.

Now try this

It does not matter which units of mass are used, as long as they are the same.

1 Iron is extracted by reducing iron(III) oxide with carbon. The theoretical yield in a particular extraction is 400 tonnes, but only 360 tonnes is obtained. Calculate the percentage yield. **(1 mark)**

2 Give three reasons why the actual yield may be less than the theoretical yield. **(3 marks)**

Atom economy

Atom economy is a way of measuring the number of atoms wasted when making a substance.

Calculating atom economy

You can calculate **atom economy** using this equation:

$$\text{atom economy} = \frac{\text{total } M_r \text{ of desired products}}{\text{total } M_r \text{ of all products}} \times 100$$

LEARN IT! IT'S NOT ON THE EQUATIONS LIST

To calculate atom economy you need to know:
- the balanced equation for the reaction
- the relative formula masses, M_r, of the products.

Remember that you can calculate the M_r of a substance by adding together the relative atomic masses, A_r, for all the atoms in its formula. You can revise this on page 19.

Worked example

Hydrogen is manufactured by the reaction of methane, found in natural gas, with steam:

$$CH_4(g) + 2H_2O(g) \rightarrow CO_2(g) + 4H_2(g)$$

Calculate the atom economy for this process. Give your answer to three significant figures. (Relative atomic masses: H = 1, C = 12, O = 16) **(2 marks)**

M_r of CO_2 = 12 + (2 × 16) = 12 + 32 = 44

M_r of H_2 = 2 × 1 = 2

total M_r of desired products = 4 × 2 = 8

total M_r of all products = 8 + 44 = 52

$$\text{atom economy} = \frac{8}{52} \times 100$$
$$= 15.4\%$$

There are two products, carbon dioxide and hydrogen, but carbon dioxide is a by-product. Hydrogen is the desired product here.

Carbon dioxide is useful for putting the fizz into fizzy drinks and filling fire extinguishers.

If the carbon dioxide is used rather than wasted, it becomes a desired product. This improves the overall atom economy of the process.

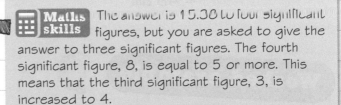 **Maths skills** The answer is 15.38 to four significant figures, but you are asked to give the answer to three significant figures. The fourth significant figure, 8, is equal to 5 or more. This means that the third significant figure, 3, is increased to 4.

100% atom economy

The atom economy of a process is 100% if all atoms in the reactants can be converted to the desired products. This happens when:

1 there is only one product, such as in the manufacture of ammonia:

$$N_2(g) + 3H_2(g) \rightleftharpoons 2NH_3(g)$$

2 all of the by-products are used, e.g. as feedstock (reactants) for other processes.

Industrial processes

In general, the higher the atom economy the lower its impact on the environment. Processes with high atom economies are usually preferred because they:

☑ produce less waste

☑ conserve limited resources.

Such processes are more **sustainable**. This means that they allow us a good standard of living today, without reducing the chances of this in the future.

Now try this

1 (a) Hydrogen can be manufactured by reacting coal with steam: $C + 2H_2O \rightarrow CO_2 + 2H_2$
   Calculate the atom economy of this process. **(3 marks)**

(b) Hydrogen can also be manufactured by the electrolysis of water: $2H_2O \rightarrow 2H_2 + O_2$
   (i) Calculate the atom economy of this process. **(3 marks)**
   (ii) Describe how the atom economy could be improved. **(1 mark)**

Molar gas volume

Avogadro's law states that equal volumes of all gases, at the same temperature and pressure, have the same number of molecules. This means that one mole of any gas occupies the same volume at a given temperature and pressure.

Gas volumes

The volume occupied by a gas depends on:
- the number of particles present
- the temperature of the gas
- the pressure of the gas.

The **molar volume** is the volume occupied by one mole of *any* gas at room temperature and atmospheric pressure (rtp).

Value for molar gas volume

The value for the molar volume, V_m, is:
- ✓ 24 dm^3/mol or
- ✓ 24 000 cm^3/mol.

You will be given these values if you need them for calculations.

Calculating the volume of gas

You can calculate the volume of a gas, if you know its amount in moles:

volume of gas = molar volume × amount

For example, to calculate the volume of 2.0 mol of CO_2 at rtp:

volume of CO_2 = 24 dm^3/mol × 2.0 mol
= 48 dm^3

Calculating the amount of gas

You can rearrange the equation to calculate the amount, in moles, of a gas:

$$amount = \frac{volume\ of\ gas}{molar\ volume}$$

LEARN IT! ITS NOT ON THE EQUATIONS LIST

For example, to calculate the amount of O_2 in 12 dm^3 of oxygen gas at rtp:

$$amount\ of\ O_2 = \frac{12\ dm^3}{24\ dm^3/mol} = 0.50\ mol$$

Worked example

Hydrogen reacts with chlorine to produce hydrogen chloride gas:

$$H_2(g) + Cl_2(g) \rightarrow 2HCl(g)$$

100 cm^3 of hydrogen reacts completely in excess chlorine. What is the maximum volume of hydrogen chloride formed at the same temperature and pressure?   **(1 mark)**

- ☐ **A** 50 cm^3
- ☐ **B** 100 cm^3
- ☒ **C** 200 cm^3
- ☐ **D** 400 cm^3

You do not have to calculate the amount of hydrogen in 100 cm^3 to answer the question. You just use ratios from the balanced equation.

The balanced equation shows that 1 mol of H_2 produces 2 mol of HCl, so:
- 100 cm^3 of H_2 produces 200 cm^3 (2 × 100 cm^3) of HCl.

It also shows that 1 mol of H_2 reacts with 1 mol of Cl_2, so:
- 100 cm^3 of H_2 reacts with 100 cm^3 of Cl_2
- if the volume of Cl_2 is more than 100 cm^3, it is in excess and unreacted chlorine will be left.

Now try this

1 Calculate the volume, in cm^3, of 0.25 mol of xenon at rtp. (V_m = 24 000 cm^3/mol)   **(1 mark)**

2 Calculate the amount of helium atoms, in mol, in 1.2 dm^3 of helium. (V_m = 24 dm^3/mol)   **(1 mark)**

3 Nitrogen reacts with hydrogen to produce ammonia: $N_2(g) + 3H_2(g) \rightarrow 2NH_3(g)$ Calculate the minimum volume, in dm^3, of hydrogen needed to make 40 dm^3 of ammonia.   **(2 marks)**

You do not need to calculate the amount in mol of these gases. Look carefully at the ratios in the balanced equation.

Gas calculations

You must be able to use molar volume in calculations involving mass and volume.

🖩 Maths skills   Changing the subject

There are two equations you need to remember for the calculations on this page:

LEARN IT!
IT'S NOT ON THE EQUATIONS LIST

- mass = M_r × amount
- volume of gas = molar volume × amount

You can rearrange the equations on the left to find the amount, in mol, of a substance:

$$\text{amount} = \frac{\text{mass}}{M_r}$$

$$\text{amount} = \frac{\text{volume of gas}}{\text{molar volume}}$$

Worked example

Sulfur burns in air: $S(s) + O_2(g) \rightarrow SO_2(g)$

Calculate the maximum volume, at room temperature and pressure, of sulfur dioxide that can be produced from 8.0 g of sulfur. (V_m = 24 dm³/mol, relative atomic mass of S = 32)

(3 marks)

$$\text{amount of S} = \frac{\text{mass}}{A_r} = \frac{8.0}{32}$$
$$= 0.25 \text{ mol}$$

From the equation, 1 mol of S makes 1 mol of SO_2, so 0.25 mol of S makes 0.25 mol of SO_2.

volume of SO_2 = molar volume × amount of SO_2
$$= 24 \times 0.25 = 6.0 \text{ dm}^3$$

These are the steps used:

1. You are told two things about S (its mass and A_r) so calculate its amount in mol.

2. Use the answer to step 1, and the balanced equation, to calculate the amount of SO_2.

3. With step 2, you now know two things about SO_2 (its amount and the molar volume) so calculate its volume.

Worked example

Lithium hydroxide is used in spacecraft to absorb carbon dioxide produced by astronauts:

$$2LiOH(s) + CO_2(g) \rightarrow Li_2CO_3(s) + H_2O(l)$$

Calculate the mass in g of lithium hydroxide needed to absorb 540 dm³ of carbon dioxide at room temperature and pressure. (V_m = 24 dm³/mol, relative formula mass of LiOH = 24) **(3 marks)**

$$\text{amount of } CO_2 = \frac{\text{volume of } CO_2}{\text{molar volume}} = \frac{540}{24}$$
$$= 22.5 \text{ mol}$$

From the equation, 1 mol of CO_2 reacts with 2 mol of LiOH, so 22.5 mol of CO_2 reacts with (2 × 22.5) = 45 mol of LiOH

mass of LiOH = M_r × amount of LiOH
$$= 24 \times 45 = 1080 \text{ g}$$

It may help you if you underline, in the equation, the substances mentioned:

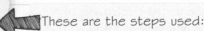

This helps you to focus on the substances you need to know about to answer the question. You can ignore the other ones, such as Li_2CO_3 and H_2O here. Note that 2 mol of LiOH reacts with 1 mol of CO_2.

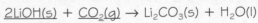

These are the steps used:

1. You are told two things about CO_2 (its volume and molar volume) so calculate its amount.

2. Use the answer to step 1, and the balanced equation, to calculate the amount of LiOH.

3. With step 2, you now know two things about LiOH (its amount and M_r) so calculate its mass.

Now try this

1. Sodium reacts with water: $2Na(s) + 2H_2O(l) \rightarrow 2NaOH(aq) + H_2(g)$

   Calculate the mass of sodium needed to produce 120 cm³ of hydrogen at rtp. (V_m = 24 000 cm³, relative atomic mass of Na = 23)

   (3 marks)

Exam skills – Chemical calculations

There may be opportunities to carry out calculations involving different aspects of chemistry. You can revise the topics for this question, which is about **chemical calculations**, on pages 60–63.

Worked example

1. Magnesium chloride can be made by reacting magnesium with dilute hydrochloric acid:

   $$Mg(s) + 2HCl(aq) \rightarrow MgCl_2(aq) + H_2(g)$$

   0.84 g of magnesium powder was added to 120 cm³ of 0.50 mol dm⁻³ hydrochloric acid.

(a) Calculate the amount, in mol, of HCl in the hydrochloric acid used.   **(2 marks)**

$$\text{volume of acid} = \frac{120}{1000} = 0.12 \text{ dm}^3$$

$$\text{amount of HCl} = 0.50 \text{ mol dm}^{-3} \times 0.12 \text{ dm}^3$$
$$= 0.060 \text{ mol}$$

> Remember to divide by 1000 to convert from cm³ to dm³.
> • amount = concentration × volume

(b) Calculate the amount, in mol, of magnesium used.   **(1 mark)**

$$\text{amount} = \frac{\text{mass of Mg}}{A_r \text{ of Mg}} = \frac{0.84}{24}$$
$$= 0.035 \text{ mol}$$

> Remember that you can find A_r values in the periodic table.
> • mass = A_r × mol
> • mass = M_r × mol

(c) Use your answers to (a) and (b) to explain which reactant was in excess.   **(2 marks)**

From the balanced equation, 1 mol of Mg reacts with 2 mol of HCl. So 0.035 mol of Mg will react with 0.070 mol of HCl. The acid here contains 0.060 mol, which is less than 0.070 mol, so the magnesium is in excess.

> If the magnesium is in excess, some of it will remain when the reaction is complete. The excess magnesium could be filtered off to leave magnesium chloride solution.

(d) The theoretical yield of magnesium chloride in this experiment is 2.85 g. The actual yield is 2.25 g. Calculate the percentage yield.   **(1 mark)**

$$\text{percentage yield} = \frac{2.25}{2.85} \times 100 = 78.9\%$$

> The percentage yield will be less than 100% because of losses. These losses can occur during filtration when some magnesium chloride solution soaks into the paper. They can also occur during heating of the solution if some spits out of the evaporating basin.

Now try this

1. Chlorine can be made by the electrolysis of concentrated sodium chloride solution.
   Overall:
   $$2NaCl + 2H_2O \rightarrow 2NaOH + H_2 + Cl_2$$
   11.7 g of sodium chloride was dissolved in 100 cm³ of water.

   (a) Calculate the concentration of sodium chloride in g dm⁻³.   **(2 marks)**

   (b) Use your answer to (a) to calculate the concentration of sodium chloride in mol dm⁻³.   **(2 marks)**

   (c) Calculate the amount, in mol, of sodium chloride in the solution.   **(2 marks)**

   (d) Use your answer to (c) to calculate the maximum amount, in mol, of chlorine that could be produced.   **(2 marks)**

   (e) Use your answer to (d) to calculate the theoretical yield, in g, of chlorine.   **(2 marks)**

   (f) Calculate the atom economy for manufacturing chlorine using this process.   **(3 marks)**

   (Use these relative atomic masses: H = 1, O = 16, Na = 23, Cl = 35.5.)

The Haber process

Reversible reactions

Chemical reactions are **reversible**. The direction of some reversible reactions can be altered by changing the reaction conditions.

Ammonium chloride decomposes when heated: $NH_4Cl(s) \rightarrow NH_3(g) + HCl(g)$

Ammonia and hydrogen chloride combine when cool:
$NH_3(g) + HCl(g) \rightarrow NH_4Cl(s)$

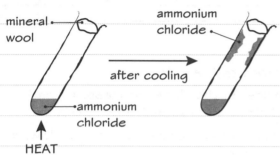

mineral wool

ammonium chloride

after cooling

ammonium chloride

HEAT

These changes can be modelled using $\rightleftharpoons$, the **reversible symbol**:

$$NH_4Cl(s) \rightleftharpoons NH_3(g) + HCl(g)$$

Dynamic equilibrium

In a **closed system**, a container where no reacting substances can enter or leave, a reversible reaction can reach **equilibrium**. At equilibrium:

☑   rate of forward reaction = rate of backward reaction

☑   the concentrations of the reacting substances stay constant (do not change).

A chemical equilibrium is a **dynamic** equilibrium:

☑   the forward and backward reactions keep going – they do not stop at equilibrium.

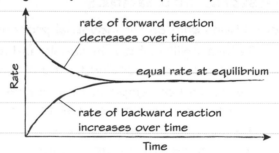

rate of forward reaction decreases over time

equal rate at equilibrium

Rate

rate of backward reaction increases over time

Time

The Haber process

The **Haber process** is a reversible reaction between nitrogen and hydrogen to form ammonia.

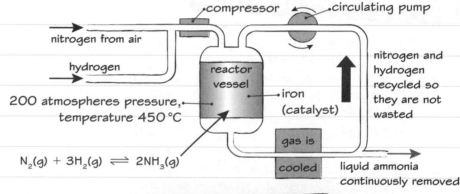

compressor   circulating pump

nitrogen from air

hydrogen

reactor vessel

iron (catalyst)

nitrogen and hydrogen recycled so they are not wasted

200 atmospheres pressure, temperature 450 °C

gas is cooled

liquid ammonia continuously removed

$$N_2(g) + 3H_2(g) \rightleftharpoons 2NH_3(g)$$

Nitrogen can be obtained by fractional distillation of liquefied air.

Methane in natural gas reacts with steam to produce carbon dioxide and hydrogen:

$$CH_4(g) + 2H_2O(g) \rightarrow CO_2(g) + 4H_2(g)$$

Worked example

The Haber process requires nitrogen and hydrogen. State the raw materials needed to obtain these two reactants. **(2 marks)**

Nitrogen is extracted from the air and hydrogen is obtained from natural gas.

Now try this

1 Hydrated copper sulfate crystals turn from blue to white when they are heated:

    hydrated copper sulfate $\rightleftharpoons$ anhydrous copper sulfate + water

        (blue)   (white)

(a) State why the word equation contains the symbol $\rightleftharpoons$. **(1 mark)**

(b) Describe how anhydrous copper sulfate could be used to test for water. **(2 marks)**

More about equilibria

The position of an equilibrium and how fast equilibrium is attained depend on the conditions.

Change in conditions	Effect on equilibrium position	Effect on rate of reaching equilibrium
temperature increased	moves in the direction of the endothermic reaction	rate increased
pressure increased	moves in the direction of the fewest molecules of gas	rate increased (if reacting gases are present)
concentration of a reacting substance increased	moves in the direction away from the reacting substance	rate increased
catalyst added	no change	rate increased

Industrial conditions

The conditions chosen for industrial processes are related to the following:

1 The availability of raw materials and energy supplies. In the Haber process, air and water are easily obtained; natural gas is more difficult and expensive to obtain.

2 The control of temperature and pressure. Many industrial processes are not allowed to reach equilibrium because this would take too long.

3 The use of a suitable catalyst, e.g. iron in the Haber process. Processes work at lower temperatures with catalysts, reducing cost and increasing yield.

Worked example

Ammonia is manufactured in the Haber process:

$$N_2(g) + 3H_2(g) \rightleftharpoons 2NH_3(g)$$

The graph shows the equilibrium yield of ammonia at different temperatures and pressures.

(a) Explain why the temperature chosen, 450 °C, is a compromise. **(3 marks)**

As the temperature increases, the equilibrium yield of ammonia decreases, but the rate of reaction increases. 450 °C is low enough to obtain an acceptable yield of ammonia, but high enough to obtain it in an acceptable time.

(b) Explain why the pressure chosen, 200 atmospheres, is a compromise. **(2 marks)**

As the pressure increases, the equilibrium yield of ammonia and the rate of reaction increase. However, very high pressures need stronger and more expensive equipment which uses more energy.

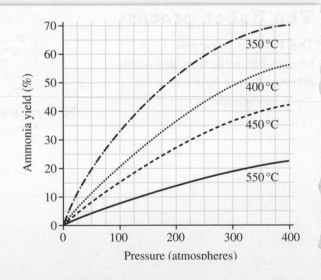

There are (1 + 3) = 4 molecules of gas on the left of the equation but only 2 on the right.

Now try this

The forward reaction is exothermic.

1 Methanol is manufactured using this reaction:
$$CO(g) + 2H_2(g) \rightleftharpoons CH_3OH(g)$$
$$\Delta H = -91 \text{ kJ mol}^{-1}$$
Predict and explain the effects on the equilibrium concentration of methanol vapour when:

(a) The pressure is decreased. **(2 marks)**
(b) The temperature is decreased. **(2 marks)**
(c) The concentration of hydrogen is increased. **(2 marks)**

Making fertilisers

Fertilisers are soluble substances added to the soil to promote the growth of plants.

NPK

Plants need many substances to grow well, but you need to know these three:
- nitrogen, N
- phosphorus, P
- potassium, K.

'NPK' fertilisers supply these elements. The elements must be supplied in soluble compounds for plant roots to absorb:

Element	Typical soluble compound
nitrogen	nitrate and ammonium salts
phosphorus	phosphate salts
potassium	potassium salts

Ammonium compounds

Ammonium salts are useful as fertilisers. They are a source of soluble **nitrogen**:

Compound	Nitrogen (%)
ammonium nitrate, NH_4NO_3	35
ammonium sulfate, $(NH_4)_2SO_4$	21

These compounds can be made by reacting ammonia solution with the appropriate acid:

$$NH_3(aq) + HNO_3(aq) \rightarrow NH_4NO_3(aq)$$

$$2NH_3(aq) + H_2SO_4(aq) \rightarrow (NH_4)_2SO_4(aq)$$

Fertiliser in the lab

Laboratory-scale production of ammonium sulfate needs:
- ✓ ammonia solution
- ✓ dilute sulfuric acid.

These are bought from chemical manufacturers. Laboratory production involves:
- ✓ only a few stages (titration then crystallisation)
- ✓ small-scale production (very little is made).

You can revise the production of pure, dry samples of soluble salts on page 39.

Fertiliser in a factory

Industrial-scale production of ammonium sulfate needs:
- ✓ natural gas, air and water (raw materials to make ammonia – see page 67)
- ✓ sulfur, air and water (raw materials to make sulfuric acid).

Industrial production involves:
- ✓ many stages (making ammonia and sulfuric acid, reacting accurate volumes, followed by evaporation)
- ✓ large-scale production (a lot is made).

Worked example

Ammonia is used in the manufacture of fertilisers.

(a) Name the acid needed to make ammonium phosphate. **(1 mark)**

Phosphoric acid.

(b) Suggest why ammonia is made using a continuous process rather than a batch process. **(1 mark)**

Ammonia is made on a large scale.

- Nitric acid makes nitrates.
- Sulfuric acid makes sulfates.
- Hydrochloric acid makes chlorides.

Laboratory production is a **batch** process:
- Small amounts are made one at a time.

The industrial production of a substance is often a **continuous** process:
- Large amounts are made all the time.

Now try this

1 (a) State the starting materials needed to produce ammonium sulfate on:
  (i) a laboratory scale **(2 marks)**
  (ii) an industrial scale **(4 marks)**

(b) Other than starting materials, describe two differences between the production of ammonium sulfate in the laboratory and production in a factory. **(2 marks)**

Fuel cells

A voltage can be produced using reactions in chemical cells and fuel cells.

🧪 Practical skills — Chemical cells

You can make a simple **chemical cell** using:

- ✓ a piece of zinc dipped in zinc sulfate solution
- ✓ a piece of copper dipped in copper sulfate solution
- ✓ a piece of filter paper soaked in potassium chloride solution
- ✓ crocodile clips, wires and a voltmeter.

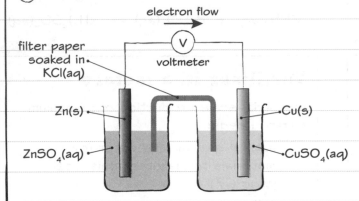

Fuel cells

In a hydrogen–oxygen **fuel cell**:

- hydrogen and oxygen are used to produce a voltage
- water vapour is the only product
- overall: $2H_2(g) + O_2(g) \rightarrow 2H_2O(g)$

This is not combustion (no flames are made).

Electrons flow through from zinc to copper (because zinc is more reactive than copper).

Ions pass through the filter paper from one beaker to the other, completing the circuit.

The ordinary batteries you use in torches and clocks are chemical cells, but the solutions are mixed with powders to make a paste.

Advantages and disadvantages of cells

Chemical cells:

- 👍 are suitable for portable appliances
- 👍 are cheap to manufacture
- 👎 may contain harmful substances
- 👎 do not produce a voltage when one of the reactants is used up – they go 'flat'.

Hydrogen–oxygen fuel cells:

- 👍 produce water as the only waste
- 👍 produce a voltage as long as the fuel and air are supplied
- 👎 are not suitable for portable appliances
- 👎 are expensive to manufacture.

Worked example

Electric cars may be powered using rechargeable chemical cells or using hydrogen–oxygen fuel cells.

Describe **one** strength and **one** weakness of using fuel cells for cars. **(2 marks)**

A fuel cell produces electricity as long as hydrogen is supplied. However, there are very few hydrogen filling stations so the car may not have a large range.

You need to be able to evaluate the strengths and weaknesses of fuel cells for given uses. Fuel cells are also used in, e.g.
- spacecraft
- submarines
- buses
- back-up electricity generators.

Now try this

1 Vehicles powered by fuel cells are very quiet.
Describe **one** possible advantage and **one** possible disadvantage of this property. **(2 marks)**

Extended response
Reversible reactions

There will be one or more 6-mark question on your exam paper. For these questions, you will need to think scientifically and structure your answer logically, showing how the points you make are related to each other.

You can revise the topics for this question, which is about **reversible reactions** and **equilibria**, on pages 67–69.

Worked example

Nitrogen and hydrogen react together to form ammonia:

$$N_2(g) + 3H_2(g) \rightleftharpoons 2NH_3(g)$$

The forward reaction is exothermic. Some ammonia would form eventually if nitrogen and hydrogen were reacted, without a catalyst, at 250 °C and 100 atmospheres pressure.

In the Haber process, a temperature of 450 °C and a pressure of 200 atmospheres are used, in the presence of an iron catalyst.

Explain why the Haber process conditions are better for the manufacture of ammonia. Give reasons for your answers.   **(6 marks)**

The catalyst does not change the equilibrium position or the equilibrium yield. However, it allows the reaction to reach equilibrium more quickly. It does this by increasing the rates of both the forward reaction and the backward reaction.

The use of a higher temperature will allow equilibrium to be reached more quickly. This means that the equilibrium yield of ammonia will be lower. If the forward reaction is exothermic, the backward reaction is endothermic. As the temperature increases, the position of equilibrium moves to the left and more ammonia breaks down.

The use of a higher pressure will also allow equilibrium to be reached more quickly. This will give a higher equilibrium yield of ammonia. There are fewer molecules of gas on the right. This means that, as the pressure increases, the position of equilibrium moves to the right.

The balanced equation provides information useful to you when you are thinking about your answer:
- The reversible symbol, $\rightleftharpoons$, shows that your knowledge and understanding of reversible reactions and equilibria may be needed.
- There are different numbers of molecules of reacting gases on each side, so the pressure will affect the position of equilibrium.
- You are told that the forward reaction is exothermic.

Notice that the Haber process uses higher pressures and temperatures than the ones mentioned first.

You should mention the **equilibrium** yield, not just the yield. Many industrial reactions are not allowed to reach equilibrium. This is to achieve an acceptable yield in an acceptable time.

You should predict and explain the effect of increasing the temperature on both the equilibrium yield and the rate of reaction.

You could also give reasons for the change in rate of reaction in terms of activation energy. You can revise this topic on pages 78 and 79.

You could also give reasons for the change in rate of reaction in terms of the frequency of collisions between reacting particles.

Your answer should show comprehensive knowledge and understanding of the topic covered. You should use the information given to you to support your explanations. Organise your explanations in a structured way with clear lines of reasoning.

Now try this

You do not need to specify values for temperature and pressure, just high or low.

Hydrogen for the Haber process can be manufactured by the reaction of methane with steam:

$$CH_4(g) + H_2O(g) \rightleftharpoons CO(g) + 3H_2(g).$$

The forward reaction is endothermic.
Predict the reaction conditions necessary to produce an acceptable yield of hydrogen in an acceptable time, and explain your answers.   **(6 marks)**

The alkali metals

The elements in group 1 of the periodic table are known as the **alkali metals**.

Physical properties

The alkali metals have some properties typical of metals. For example, they are:
- good conductors of heat and electricity
- shiny when freshly cut.

However, compared with most other metals, they:
- are soft (you can cut them with a knife)
- have relatively low melting points (but all are solid).

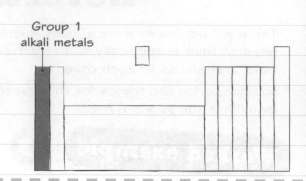

Group 1
alkali metals

Reactions with water

The alkali metals react with water, producing an alkaline metal hydroxide and hydrogen, e.g.

sodium + water → sodium hydroxide + hydrogen

$$2Na(s) + 2H_2O(l) → 2NaOH(aq) + H_2(g)$$

Reactivity

The **reactivity** of the alkali metals increases down the group:
- Lithium fizzes steadily.
- Sodium melts into a ball from the heat released in the reaction and fizzes rapidly.
- Potassium gives off sparks and the hydrogen produced burns with a lilac-coloured flame.

Density

Lithium, sodium and potassium are less dense than water, so they float.

Storage

Lithium, sodium and potassium are stored in oil. This is to keep air and water away.

Worked example

(a) Rubidium is placed below potassium in the periodic table.
Predict what is seen in the reaction of rubidium with water. **(2 marks)**

Rubidium will react very vigorously with water, producing sparks and bursting into flames explosively.

(b) Explain why sodium is more reactive than lithium. **(2 marks)**

Sodium atoms are larger than lithium atoms. So the outer electron in a sodium atom is further from the nucleus than the outer electron in a lithium atom. This means that the force of attraction is weaker. So the outer electron is lost more easily from sodium than from lithium.

You need to be able to:
- describe the pattern of reactivity of lithium, sodium and potassium, and
- use it to predict the reactivity of other alkali metals (rubidium, caesium and francium).

The alkali metals all have one electron in the outer shell of their atoms. They lose this electron in reactions to form ions with a 1+ charge, e.g. Li^+ and Na^+. The more easily the outer electron is lost, the more reactive the metal.

Going down the group, the number of occupied shells in an atom increases. The electrons in the inner shells **shield** the outer electron, reducing its attraction for the nucleus.

Now try this

1 Caesium, Cs, is placed below rubidium in group 1 of the periodic table.
(a) Write a balanced equation for the reaction of caesium with water. **(2 marks)**

(b) Predict what is seen in the reaction of caesium with water. **(2 marks)**

(c) State **two** physical properties that rubidium and caesium have in common. **(2 marks)**

The halogens

The elements in group 7 of the periodic table are non-metals known as the **halogens**.

Appearance

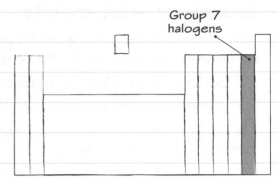
Group 7 halogens

Element	State at room temperature	Colour
fluorine, F_2	gas	pale yellow
chlorine, Cl_2	gas	yellow–green
bromine, Br_2	liquid	red–brown
iodine, I_2	solid	dark grey

covalent molecules, each with two atoms

forms a purple vapour when warmed

Melting and boiling points

Going down group 7:
- melting points increase
- boiling points increase.

When simple molecular substances melt or boil:
- weak intermolecular forces are overcome
- the strong covalent bonds joining atoms together in each molecule do not break.

Going down group 7:
- the intermolecular forces between molecules become stronger
- more heat energy is needed to overcome these forces.

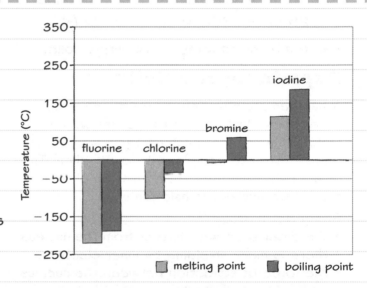

You can revise simple molecular substances on page 14 and physical states on page 26.

Worked example

Chlorine is manufactured on an industrial scale by the electrolysis of concentrated sodium chloride solution.

Describe a simple laboratory test for the presence of chlorine. **(2 marks)**

Put a piece of damp blue litmus paper into the container. If chlorine is present, the litmus paper turns red, then is bleached white.

Practical skills When you describe a laboratory test, say what you would do and what you would observe.

In a different test, damp **starch iodide paper** turns blue–black in the presence of chlorine. The colour change happens because chlorine displaces iodine, which then reacts with the starch. You can revise halogen displacement reactions on page 75.

Now try this

Use evidence from the bar chart to explain your answer.

1 Astatine is a rare, radioactive element placed immediately below iodine in group 7.

   (a) Suggest why astatine is thought to be dark in colour, even though no one has seen it. **(1 mark)**

   (b) Predict the melting point of astatine, and justify your answer. **(2 marks)**

Reactions of halogens

The reactivity of the elements decreases down group 7 (the opposite trend to group 1).

Reactions with metals

The halogens react with metals to produce compounds called metal halides, e.g.

sodium + chlorine → sodium chloride

$$2Na(s) + Cl_2(g) \rightarrow 2NaCl(s)$$

The sodium burns with an orange flame, forming white sodium chloride.

Iron wool reacts with the halogens when it is heated, e.g.

iron + chlorine → iron(III) chloride

$$2Fe(s) + 3Cl_2(g) \rightarrow 2FeCl_3(s)$$

The iron burns to form a dark-purple solid.

Halide ions, halides

The elements in group 7 are called **halogens**. In reactions with metals, halogen atoms gain electrons and are reduced, e.g.

$$Cl_2 + 2e^- \rightarrow 2Cl^-$$

The ions formed:

☑ have a 1– charge, e.g. Cl^-, Br^- and I^-

☑ are called **halide ions**.

Halides are compounds of metals or hydrogen with halogens, such as sodium chloride.

Explaining reactivity

A halogen atom has seven electrons in its outer shell. When a halogen reacts with a metal or hydrogen, each halogen atom gains one electron to complete its outer shell. The less easily a halogen atom gains an electron, the less reactive the halogen is.

Going down group 7:
• the outer shell gets further from the nucleus
• there is more shielding by inner electrons
• the force of attraction between the nucleus and outer shell electrons gets weaker
• electrons are gained less easily
• the elements become less reactive.

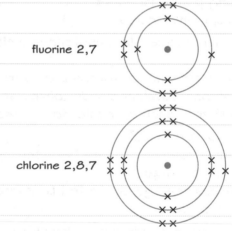

fluorine 2,7

chlorine 2,8,7

Fluorine is more reactive than chlorine – its atoms gain electrons more easily.

Worked example

Hydrogen reacts vigorously with bromine vapour to produce hydrogen bromide gas:

$$H_2(g) + Br_2(g) \rightarrow 2HBr(g)$$

What happens if this gas is added to water? **(1 mark)**

☐ A It does not dissolve.

☐ B It dissolves to form a neutral solution.

☐ C It dissolves to form an alkaline solution.

☒ D It dissolves to form an acidic solution.

Hydrogen reacts with the halogens to produce hydrogen halides. The reactions become less vigorous down group 7:
• fluorine reacts explosively in the dark
• chlorine reacts explosively in sunlight
• bromine reacts vigorously in a flame.

The hydrogen halides all dissolve in water to produce acidic solutions, e.g.
• hydrogen chloride gas, HCl(g), produces hydrochloric acid, HCl(aq). This fully dissociates to form $H^+(aq)$ and $Cl^-(aq)$ ions.

Now try this

1 Cold iron wool burns in fluorine but hot iron wool reacts slowly with iodine vapour. Explain why iodine is less reactive than fluorine. **(2 marks)**

2 Aluminium powder reacts vigorously with iodine to produce aluminium iodide, AlI_3. Write a balanced equation for the reaction. **(2 marks)**

Halogen displacement reactions

A more reactive halogen can displace a less reactive halogen from its compounds.

Practical skills — Investigating displacement

You can demonstrate **displacement reactions** by adding a halogen solution to a metal halide solution, then seeing if the mixture darkens. For example:

✓ Chlorine in 'chlorine water' displaces bromine from aqueous sodium bromide solution.

✓ balanced equation: $Cl_2(aq) + 2NaBr(aq) \rightarrow 2NaCl(aq) + Br_2(aq)$

✓ ionic equation (Na^+ ions are spectator ions – see page 3):

$Cl_2(aq) + 2Br^-(aq) \rightarrow 2Cl^-(aq) + Br_2(aq)$

chlorine water

the reddish-brown colour is due to the bromine that has been displaced

sodium chloride solution + bromine

sodium bromide solution

Worked example

A student adds a few drops of halogen solutions to small volumes of potassium halide solutions on a spotting tile. Put a tick (✓) in the table below to show where she sees the mixture turning darker. **(2 marks)**

	Chloride	Bromide	Iodide
Chlorine	not done	✓	✓
Bromine		not done	✓
Iodine			not done

> The halogens are toxic. It is safer to use small volumes of dilute halogen solutions.

> Potassium salts form colourless solutions. They darken if bromine or iodine is produced. A halogen cannot displace itself so, for example, chlorine water is not added to potassium chloride solution in the investigation.

Redox reactions

Halogen displacement reactions are **redox** reactions, e.g. when chlorine displaces bromine from bromide ions in solution:

1 Chlorine atoms gain electrons and are reduced to chloride ions:

$Cl_2(aq) + 2e^- \rightarrow 2Cl^-(aq)$

2 Bromide ions lose electrons and are oxidised to bromine:

$2Br^-(aq) \rightarrow Br_2(aq) + 2e^-$

OIL RIG

✓ **O**xidation **i**s **l**oss of electrons

✓ **R**eduction **i**s **g**ain of electrons.

These processes happen together, at the same time, in **redox** reactions.

Metal displacement reactions are also redox reactions. You can revise these reactions on page 43.

Now try this

> Astatine, At, is at the bottom of group 7.

1 Aqueous bromine solution reacts with aqueous potassium iodide solution.
   (a) For this reaction, write the balanced equation and the ionic equation. **(4 marks)**

(b) Explain, with the help of half-equations, why this reaction is a redox reaction. **(4 marks)**

(c) State and explain whether astatine will react with potassium iodide solution. **(2 marks)**

The noble gases

The elements in group O of the periodic table are known as the **noble gases**.

Chemical properties

The noble gases are chemically **inert**. Their lack of reactivity is because:

- their atoms have full outer shells of electrons, so
- they have no tendency to lose, gain or share electrons.

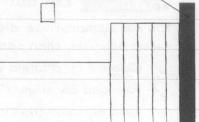

Group O
noble gases

Element	He	Ne	Ar
Electronic configuration	2	2.8	2.8.8

Uses of the noble gases

You need to be able to explain how use of a noble gas depends on a particular property. The table shows some typical uses of noble gases.

Noble gas	Use	Property needed		Reason for use
		Low density	Inertness	
helium	**lifting gas** in party balloons and airships	✓	✓	• helium is less dense than air so the balloons and airships rise • it is non-flammable so does not ignite
argon krypton xenon	**filling gas** in filament lamps		✓	• the metal filament becomes hot enough to glow • the inert gases stop it burning away
argon	**shield gas** during welding		✓	• argon is denser than air so it keeps air away from the metal • it is inert so the metal does not oxidise

Worked example

The table shows the densities and boiling points of some of the noble gases. Describe the trends in these properties and use them to predict the missing numbers. **(4 marks)**

Element	Density (kg/m³)	Boiling point (°C)
helium, He	0.15	−269
neon, Ne	1.20	
argon, Ar		−186
krypton, Kr	2.15	−152

The densities increase as you go down the group. The boiling points get higher as you go down the group.

The density of argon could be about 1.7 or 1.8 kg/m³.

The boiling point of neon could be about −230°C.

These numbers are about halfway between the numbers above and below in the table.

You need to be able to answer questions like this one that ask you to identify a trend, and use the trend to predict missing values.

Now try this

1. The density of air is 1.225 kg/m³. Explain why helium and neon can be used as lifting gases, but argon and krypton cannot. **(2 marks)**

Look at the data in the Worked example.

2. Give **two** reasons why argon may be used in some fire-extinguishing systems. **(2 marks)**

3. Explain, in terms of its electronic configuration, why helium is inert. **(3 marks)**

Extended response – Groups

There will be one or more 6-mark question on your exam paper. For these questions, you will need to think scientifically and structure your answer logically, showing how the points you make are related to each other.

You can revise the topics for this question, which is about **groups in the periodic table**, on pages 72–76.

Worked example

Lithium and potassium react with cold water. Similar products form in these reactions.
Compare and contrast the way in which these two elements react with water. Explain your answers in terms of electronic configurations, reactions and products. **(6 marks)**

Both elements are in group 1 of the periodic table, the alkali metals. Their atoms have one electron in their outer shells, which is why they react in a similar way (lithium is 2.1 and potassium is 2.8.8.1).

Both metals float on water and both produce hydrogen. They also produce soluble metal hydroxides when they react with water, forming alkaline solutions.

However, potassium is more reactive than lithium, so its reaction with water is more vigorous. This is because potassium atoms lose their outer electron more easily. Potassium atoms have more filled electron shells, so their outer electron is further from the nucleus and more shielded, so has a weaker attraction for it.

Lithium fizzes slowly when it reacts with water, gradually becoming smaller until it disappears. Potassium reacts more vigorously. It immediately ignites with a lilac flame when it reacts with water, and disappears with sparks.

Command words: Compare and contrast

The question is looking for the similarities and differences between two or more things, without needing a conclusion. The answer must relate to all things mentioned in the question. It must include at least one similarity and one difference.

 You should mention which group the elements are placed in. Their electronic configurations are important in explaining why their reactions are similar.

 You should be clear about what is similar in the reactions of lithium and potassium with water.

 Your answer should compare the reactivity of the two metals, and explain this in terms of how easily the outer electrons are lost in reactions.

 You need to be able to describe the reactions of the alkali metals with water. The key differences in the reactions of lithium and potassium are mentioned here.

Your answer should show comprehensive knowledge and understanding of the topic covered. Take care to cover both similarities and differences. Make sure that you use any information given to you to support your explanations. Organise your explanations in a structured way with clear lines of reasoning.

Now try this

The command word **devise** means that you need to plan or invent a procedure from existing principles or ideas.

The elements in group 7 of the periodic table include chlorine, bromine and iodine. Devise an experiment to determine the order of reactivity of these elements using displacement reactions.
In your answer, assume that you are given dropping pipettes, test tubes and these solutions:
- chlorine solution, bromine solution, iodine solution
- potassium chloride solution, potassium bromide solution, potassium iodide solution.

Describe suitable safety precautions. Include the expected results and explain how you would use them to show the order of reactivity. You may include balanced equations to support your answer. **(6 marks)**

Rates of reaction

The greater the frequency of successful collisions, the greater the rate of reaction.

Colliding particles

For a reaction to happen:
- reactant particles must collide with each other **and**
- the collisions must have enough energy.

The **activation energy** of a reaction is the minimum energy needed by reactant particles for a reaction to happen. A **successful collision** has the activation energy or more.

Rate and time

The greater the **rate of reaction**, the lower the **reaction time**:
- ✓ a fast reaction happens in a short time, such as combustion
- ✓ a slow reaction happens over a long time, such as rusting.

Take care not to confuse the two quantities.

Concentration and pressure

The rate of reaction increases if the **concentration** of a dissolved reactant increases, or if the **pressure** of a reacting gas increases:
- there are more particles in the same volume
- the frequency of successful collisions increases.

Surface area:volume ratio

The rate of reaction increases when the **surface area:volume ratio** of a solid reactant increases, e.g. when lumps are made into a powder:
- more particles of reactant are available
- the frequency of successful collisions increases.

Temperature

The rate of reaction increases when the **temperature** increases because the particles gain energy, and:
- the particles move faster
- the frequency of collisions increases **and**
- the energy of collisions increases, so a greater proportion of collisions are successful.

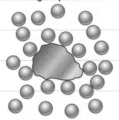

larger piece

smaller surface area
slower reaction

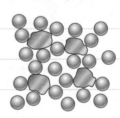

smaller pieces

larger surface area
faster reaction

Worked example

(a) Describe three features of a catalyst.

(3 marks)

Catalysts are substances that speed up the rate of a reaction without altering the products. They are unchanged chemically and in mass at the end of the reaction.

(b) Explain how a catalyst works.   **(2 marks)**

Catalysts provide an alternative reaction pathway with a lower activation energy.

Catalysts are often transition metals or their compounds, such as iron in the Haber process.

Enzymes are **biological** catalysts. Enzymes found in yeast catalyse the conversion of glucose to ethanol and carbon dioxide. Enzymes are used in the production of wine and other alcoholic drinks.

With a lower activation energy, a greater proportion of collisions have the necessary activation energy or more. So the frequency of successful collisions is greater at a given temperature.

Now try this

1 Marble chips react with hydrochloric acid: $CaCO_3(s) + 2HCl(aq) \rightarrow CaCl_2(aq) + H_2O(l) + CO_2(g)$
   State and explain the effect of the following on the rate of the reaction:

   (a) crushing the chips   **(3 marks)**
   (b) diluting the acid   **(3 marks)**

   (c) heating the acid.   **(4 marks)**

 **Practical skills**

Investigating rates

Core practical

Effect of surface area on rate

Aims

To investigate the effect on the rate of reaction with hydrochloric acid of changing the surface area of calcium carbonate.

Apparatus

- eye protection
- conical flask
- measuring cylinder
- marble chips (small, medium, large)
- cotton wool
- dilute hydrochloric acid
- ±0.01 g balance
- stop clock.

Method

1. Prepare three sets of marble chips on folded paper, one for each size of chip. Adjust the numbers so that each set has the same mass.

2. Add a measured volume of dilute acid to the conical flask. Plug the flask with cotton wool.

3. Place the flask and a set of marble chips on the balance. Record the reading.

4. Remove the cotton wool. Add the chips to the acid. Start the stop clock, and replace the cotton wool and folded paper.

5. Record the mass each 30 s for a few minutes.

Repeat steps 1 to 5 with the other sets of chips.

Results

Time (s)	Mass (g)	Change in mass (g)
0	70.00	0.00
30	69.90	0.10
60	69.83	0.17

Analysis

Calculate the change in mass at each time, t:
- change in mass = mass at start − mass at t

For each set of chips, calculate the mean rate of reaction over the same amount of time:

$$\text{mean rate of reaction} = \frac{\text{change in mass}}{\text{chosen time}}$$

You could also investigate the effect of changing the concentration of the hydrochloric acid.

You could use powdered marble. However, the rate of reaction may be so great that it becomes difficult to measure the mass at a given time.

Instead of recording the change in mass, you could use a gas syringe to measure the volume of carbon dioxide produced.

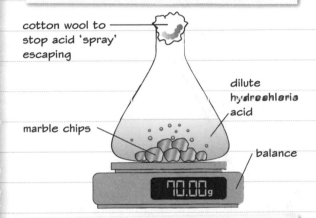

cotton wool to stop acid 'spray' escaping

dilute hydrochloric acid

marble chips

balance

You should record your results for each set of chips in a suitable table. Include a column for the change in mass at each time. Make sure that you have enough rows for all your readings.

You could also plot line graphs:
- change in mass in g on the vertical axis
- time in s on the horizontal axis
- smooth line of best fit, ignoring anomalous results.

You can revise how to interpret the results from experiments similar to this on page 80.

Now try this

A student investigates the effect of acid concentration on the rate of this reaction:

$$Na_2S_2O_3(aq) + 2HCl(aq) \rightarrow 2NaCl(aq) + H_2O(l) + SO_2(g) + S(s)$$

The reaction mixture becomes cloudy as sulfur forms. The student measures the time taken for it to become too cloudy to see a cross drawn on a piece of paper underneath the flask. In each experiment, he uses the same volume and concentration of sodium thiosulfate solution. He uses the same volume of diluted acid each time. Explain three steps that the student takes to obtain results that allow a fair comparison. **(4 marks)**

Exam skills – Rates of reaction

There may be opportunities to draw or complete graphs and to interpret them. You can revise the topics for this question, which is about **rates of reaction**, on pages 78 and 79.

Worked example

Hydrogen peroxide solution decomposes to form water and oxygen: $2H_2O_2(aq) \rightarrow 2H_2O(l) + O_2(g)$
Manganese dioxide powder, MnO_2, is a catalyst for this reaction. A student investigates the rate of the catalysed reaction. He measures the volume of oxygen produced over 2 minutes at 20 °C.

Time (s)	0	15	30	45	60	75	90	105	120
Volume of oxygen (cm³)	0	20	34	43	47	49	50	50	50

(a) Plot these results on the grid. **(3 marks)**

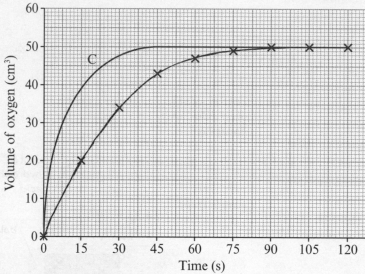

Time (s)

Command word: Plot

You need to produce a graph by marking points accurately on a grid using given data. Then draw a line of best fit through these points. You must include a suitable scale and labelled axes if these are not given to you.

You should aim to plot each point to at least half the correct square. A line of best fit need not be a straight line (it could be a smooth curve), but:

✓ draw it in one go, not in bits.

Line C, the answer to (**c**), lies to the left of the first line and ends at the same volume.

No more oxygen is produced after 90 s, so the reaction time is 90 s, not the duration of the investigation.

(b) Calculate the mean rate of reaction, giving your answer to two significant figures. **(2 marks)**

volume of oxygen produced = 50 cm³

reaction time = 90 s

mean rate = $\dfrac{\text{volume}}{\text{time}} = \dfrac{50\,\text{cm}^3}{90\,\text{s}} = 0.56$ cm³/s

(c) On the same axes, sketch the curve you would expect if the student used the same volume of hydrogen peroxide solution at 40 °C.
Label this curve **C**. **(2 marks)**

Command word: Sketch

You need to do a freehand drawing or graph. You may need labelled axes (not to scale), a line and key features identified.

Now try this

(a) Use the student's results to calculate the mean rate of reaction between 0 s and 45 s, and between 45 s and 90 s. **(4 marks)**

(b) Explain, in terms of particles and their collisions, why the mean rate of reaction differs in each of your answers to (**a**). **(4 marks)**

(c) Describe how the student could show that the mass of catalyst is unchanged in the reaction. **(4 marks)**

Heat energy changes

Temperature changes

Most reactions involve temperature change:
- In **exothermic reactions**, heat energy is given **out**, and the reaction mixture or the surroundings increase in temperature.
- In **endothermic reactions**, heat energy is taken **in**, and the reaction mixture or the surroundings decrease in temperature.

Reaction	Exothermic	Endothermic
neutralisation	✓	✓
displacement	✓	
precipitation	✓	✓
dissolving	✓	✓

Temperature can go up or down depending on the precipitate or salt.

Practical skills — Measuring temperature changes

You can use this apparatus to investigate the temperature changes in reactions.

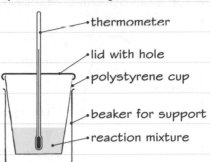

- thermometer
- lid with hole
- polystyrene cup
- beaker for support
- reaction mixture

The beaker and air inside are further insulation to reduce energy transfer.

Bonds and energy

When a chemical reaction happens, the bonds that hold the atoms together in the molecules of the reactants are broken. The atoms then come together in new arrangements to form the products.

- Breaking bonds is endothermic (energy is needed).
- Making bonds is exothermic (energy is released).

exothermic reaction

reactants → products

Energy

Overall energy is released to surroundings and this makes the reaction exothermic.

Energy is released to the surroundings because more heat energy is released making bonds in the products than is needed to break bonds in the reactants.

endothermic reaction

reactants → products

Energy

Overall energy is taken in to the reaction and this makes the reaction endothermic.

Energy is taken in from the surroundings because less heat energy is released making bonds in the products than is needed to break bonds in the reactants.

Worked example

Calcium chloride is added to water and stirred. Explain why the mixture warms up. **(3 marks)**

An exothermic change happens. More heat energy is released in forming bonds in the products than is needed to break bonds in the reactants. So, overall, heat energy is given out.

The temperature does not always go up when solutions form. For example, the temperature goes down when ammonium nitrate dissolves.

The temperature increases so this must be an exothermic change. Remember:
- exothermic = energy in or 'energy exits'
- endothermic = energy out or 'energy enters'

Now try this

1 Copy and complete the table by putting a tick (✓) in each correct box. **(4 marks)**

	Breaking bonds	Making bonds	Temperature of reaction mixture	
			Increases	Decreases
Exothermic process				
Endothermic process				

Reaction profiles

Energy profile diagrams model the energy changes that happen during reactions.

Exothermic reactions

The diagram shows a typical **reaction profile** for an exothermic reaction.

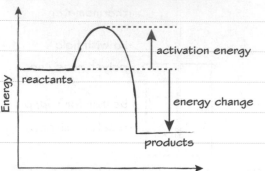

You should see that:

- ☑ the **energy level** of the reactants is greater than the energy level of the products
- ☑ the **energy change** of the reaction is negative (energy is transferred to the surroundings).

The **activation energy** is the minimum energy needed to start a reaction. It may be supplied by, for example:

- ☑ heating the reaction mixture
- ☑ applying a flame or a spark.

Endothermic reactions

The diagram shows a typical reaction profile for an endothermic reaction.

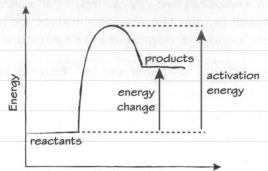

You should see that:

- ☑ the energy level of the reactants is lower than the energy level of the products
- ☑ the energy change of the reaction is positive (energy is transferred from the surroundings).

The activation energy may be supplied by, for example:

- ☑ continually heating the reaction mixture
- ☑ passing an electric current through an electrolyte (as in electrolysis).

Worked example

The reaction profile represents an exothermic reaction.

(a) Explain what the energy change shown as A represents.   **(2 marks)**

The activation energy, which means that this is the energy needed to start the reaction.

(b) Draw a line to represent the change in energy level when a catalyst is added.   **(1 mark)**

Energy ↑

A

reactants

B

products

In this diagram, B represents the overall energy change. A catalyst provides an alternative pathway with a lower activation energy for the chemical reaction.

Now try this

1 Explain the term 'activation energy'.   **(2 marks)**

2 Sketch reaction profiles for the following reactions:

(a) $CH_4 + 2O_2 \rightarrow CO_2 + 2H_2O$ (exothermic)
    (4 marks)

(b) $CaCO_3 \rightarrow CaO + CO_2$ (endothermic)
    (4 marks)

Calculating energy changes

You can calculate the energy change in a reaction using bond energies.

Breaking and making bonds

A **bond energy** is the energy needed to break 1 mol of a particular covalent bond, for example:

- 413 kJ is needed to break 1 mol of C–H bonds
- 413 kJ is released when 1 mol of C–H bonds forms.

Different bonds have different bond energies, depending on factors such as the elements involved and the length of the bond.

Some bond energies

The table shows some examples of bond energies, which are all positive.

Bond	Bond energy (kJ mol⁻¹)
C–H	413
H–H	436
O–H	464
O=O	498
C=O	805

Do not try to learn any of these values – you will be given any that you need for calculations.

Worked example

Hydrogen burns in oxygen to form water. The equation shows the structures of the substances involved in this balanced reaction.

$$2 \times (H-H) + O=O \longrightarrow 2 \times \left(\begin{smallmatrix} & O & \\ H & & H \end{smallmatrix} \right)$$

(a) Calculate the energy needed to break all the bonds in the reactants. **(3 marks)**

$2 \times (H–H) = 2 \times 436 = 872$ kJ mol⁻¹

$1 \times (O=O) = 1 \times 498 = 498$ kJ mol⁻¹

total $= 872 + 498 = 1370$ kJ mol⁻¹

(b) Calculate the energy released when new bonds form in the products. **(2 marks)**

$4 \times (O–H) = 4 \times 464 = 1856$ kJ mol⁻¹

(c) Calculate the overall energy change for the reaction. **(2 marks)**

Energy change $= 1370 - 1856 = -486$ kJ mol⁻¹

> You need to use the displayed formula of each substance to work out how many bonds of each type it contains.

> Multiply the bond energy for each type of bond by the number of those bonds present. Take care to use the correct values.
>
> Notice that O=O is the energy needed to break the double bond.

> Each water molecule has two O–H bonds. The total number of O–H bonds in two water molecules is (2 × 2) = 4.

> $$\text{energy change} = \text{total energy in} - \text{total energy out}$$
>
> The negative sign in the final answer shows that the reaction is exothermic.
>
> The energy change for an endothermic reaction is positive.

Now try this

Use the bond energies in the table at the top of the page to help you answer these questions.

1 During electrolysis, water decomposes to form hydrogen and oxygen.
   (a) Calculate the overall energy change in the reaction. **(4 marks)**
   (b) Explain whether the process is exothermic or endothermic. **(2 marks)**

 $2 \times \left(\begin{smallmatrix} & O & \\ H & & H \end{smallmatrix} \right) \longrightarrow 2 \times (H-H) + O=O$

2 Methane burns in oxygen to form carbon dioxide and water.
   Calculate the overall energy change in the reaction. **(5 marks)**

$$H-\overset{\displaystyle H}{\underset{\displaystyle H}{C}}-H + 2 \times (O=O) \longrightarrow 2 \times \left(\begin{smallmatrix} & O & \\ H & & H \end{smallmatrix} \right) + O=C=O$$

Crude oil

Crude oil is a fossil fuel, formed over millions of years from ancient remains of marine organisms.

Hydrocarbons

Hydrocarbons are compounds of carbon and hydrogen atoms **only**. Carbon atoms can form four covalent bonds. In a hydrocarbon molecule, these bonds can be:

- carbon–carbon bonds
- carbon–hydrogen bonds.

Hydrocarbon molecules can consist of:

- chains (with or without branches) or rings of carbon atoms.

Hydrocarbons in crude oil

Crude oil is:

- a complex mixture of hydrocarbons, with their carbon atoms in chains or rings
- an important source of useful substances
- a finite resource.

Finite resources

Finite resources:

✓ are no longer being made, or

✓ are being made extremely slowly.

Crude oil takes millions of years to form.

Worked example

(a) Hydrocarbons are found in crude oil. Which of the following formulae represents a hydrocarbon? **(1 mark)**

☐ **A** C_2H_5OH

☒ **B** C_2H_6

☐ **C** $C_6H_{12}O_6$

☐ **D** CCl_4

(b) Describe **two** ways in which crude oil is an important source of useful substances.

(2 marks)

Hydrocarbons from crude oil are useful as fuels and as feedstock for the petrochemical industry.

The chemical symbols you see in the formulae for hydrocarbon molecules are:

- C for carbon
- H for hydrogen.

In **ball-and-stick models** (similar to the ones above), atoms are usually modelled as:

- black for carbon atoms
- white for hydrogen atoms.

Familiar fuels such as petrol and diesel oil (used in cars) come from crude oil.

A **feedstock** is a starting material for an industrial chemical process.

The **petrochemical** industry involves the use and manufacture of substances from crude oil.

You can revise the manufacture of polymers from hydrocarbon feedstock on page 101.

Now try this

1 (a) Explain the meaning of the term 'hydrocarbon'. **(2 marks)**

   (b) State the type of chemical bond found in a hydrocarbon molecule. **(1 mark)**

2 Explain why crude oil is described as a **finite** resource. **(2 marks)**

3 Describe the possible arrangements of carbon atoms in hydrocarbon molecules.

   (2 marks)

4 (a) Give an example of a fuel obtained from crude oil. **(1 mark)**

   (b) Give an example of a substance manufactured using crude oil as a feedstock. **(1 mark)**

Fractional distillation

Fractional distillation is used to separate crude oil into simpler, more useful mixtures.

Fractional distillation

Crude oil can be separated by fractional distillation because its different hydrocarbons have different **boiling points**. During fractional distillation:

- oil is heated to evaporate it
- vapours rise in a fractionating column
- the column has **a temperature gradient** – hot at the bottom, cool at the top
- each fraction condenses where it becomes cool enough, and is piped out of the column.

The **gases** fraction does not condense and leaves at the top.

The **bitumen** fraction does not evaporate and leaves at the bottom.

The other fractions are liquid at room temperature and are useful as fuels.

Fractions

A **fraction** is a mixture of hydrocarbons with **similar** boiling points and numbers of carbon atoms. There are trends in the properties of the different fractions from crude oil:

Number of C and H atoms	Boiling point	Ease of ignition	Viscosity
smallest	lowest	most flammable	least viscous
↑	↑	↑	↑
largest	highest	least flammable	most viscous

Substances with a low viscosity are runny, and those with a high viscosity are 'thick'.

Alkanes

Most of the hydrocarbons from crude oil are **alkanes**, a **homologous series** of compounds.

You can revise alkanes and homologous series on page 86.

Boiling point and molecule size

As the number of carbon and hydrogen atoms in a hydrocarbon molecule increases:

- ✓ the strength of the **intermolecular forces** increases
- ✓ more energy must be transferred to overcome these forces
- ✓ the boiling point increases.

Worked example

Draw one straight line from the name of each fraction to its correct typical use.   **(5 marks)**

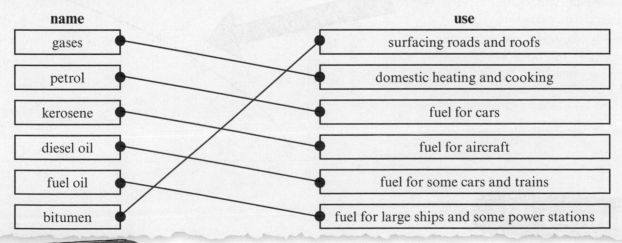

name	use
gases	surfacing roads and roofs
petrol	domestic heating and cooking
kerosene	fuel for cars
diesel oil	fuel for aircraft
fuel oil	fuel for some cars and trains
bitumen	fuel for large ships and some power stations

Now try this

1 Suggest two reasons to explain why bitumen is not suitable for use as a fuel. **(2 marks)**

 2 Describe how crude oil is separated into fractions by fractional distillation. **(3 marks)**

Alkanes

The **alkanes** are a homologous series of hydrocarbons.

Features of a homologous series

A **homologous series** is a series of compounds:
- in which **molecular formulae** of neighbouring members differ by CH_2
- that show a gradual variation in physical properties, such as boiling points
- that have similar chemical properties.

Chemical properties of alkanes

The alkanes undergo **complete combustion**. When they react completely with oxygen, carbon dioxide and water vapour form.

For example, methane (from natural gas):

methane + oxygen → carbon dioxide + water

$$CH_4(g) + 2O_2(g) \rightarrow CO_2(g) + 2H_2O(g)$$

Alkanes

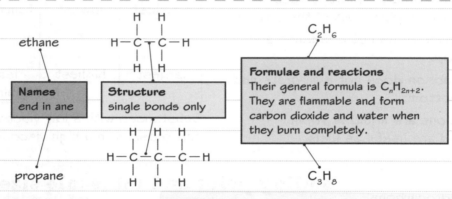

ethane

C_2H_6

Names end in ane

Structure single bonds only

Formulae and reactions
Their general formula is C_nH_{2n+2}. They are flammable and form carbon dioxide and water when they burn completely.

propane

C_3H_8

 Worked example

The molecular formula shows the number of carbon atoms (and hydrogen atoms) in the molecule.

Going from one alkane to the next, the formula changes by CH_2.

The table shows the boiling points of some alkanes.

Molecular formula of alkane	CH_4	C_2H_6	C_4H_{10}	C_6H_{14}	C_8H_{18}	$C_{10}H_{22}$	$C_{16}H_{34}$
Boiling point of alkane (°C)	−162	−89	0	69	126	174	287

Complete the graph below, and draw a line of best fit. **(3 marks)**

The boiling point increases as the number of carbon atoms in the alkane molecule increases.

This is why fractional distillation produces fractions with similar boiling points and number of carbon atoms in their molecules.

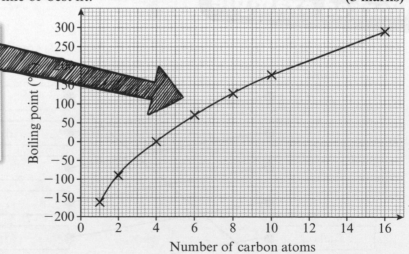

Now try this

1 State three features of a homologous series. **(3 marks)**

2 The molecular formula for icosane is $C_{20}H_{42}$. Predict the molecular formula for heneicosane, the next member of the alkane homologous series. **(1 mark)**

3 Use the Worked example graph to predict the boiling point of propane, C_3H_8. **(1 mark)**

Incomplete combustion

Incomplete combustion happens when the supply of oxygen to a burning fuel is limited.

Complete versus incomplete combustion

During **complete combustion** of a hydrocarbon fuel, such as petrol, kerosene or diesel oil:

- hydrogen is oxidised to water vapour, H_2O
- carbon is oxidised to carbon dioxide, CO_2
- energy is given out (transferred to the surroundings by radiation as heat and light).

During **incomplete combustion**, hydrogen is still oxidised to water vapour, but:

- carbon may be partially oxidised to carbon monoxide, CO
- carbon may be released as carbon particles or **soot**
- less energy is given out.

During incomplete combustion, different amounts of the different carbon products form, depending on how much oxygen is available for oxidation. For example:

$$\text{methane} + \text{oxygen} \rightarrow \text{carbon} + \text{carbon monoxide} + \text{carbon dioxide} + \text{water}$$
$$4CH_4(g) + 6O_2(g) \rightarrow C(s) + 2CO(s) + CO_2(g) + 8H_2O(g)$$

Cars and appliances

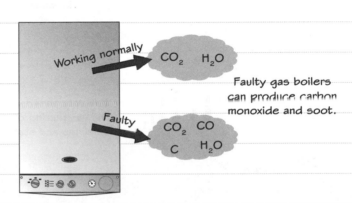

Faulty gas boilers can produce carbon monoxide and soot.

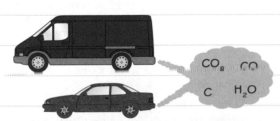

There is always incomplete combustion in vehicle engines.

Worked example

(a) Explain how carbon monoxide behaves as a toxic gas. **(2 marks)**

Carbon monoxide attaches to haemoglobin in red blood cells, preventing oxygen attaching instead. This reduces the amount of oxygen carried around the body by the bloodstream.

(b) Describe **two** problems caused by soot. **(2 marks)**

Soot can build up in chimneys where it may eventually cause fires; it also blackens buildings.

Carbon monoxide can cause unconsciousness and even death. It is:

- colourless
- odourless (has no smell).

Electronic carbon monoxide detectors are used to warn us when the gas is present.

Tiny soot particles can be breathed in. This may cause lung diseases such as bronchitis, or make existing lung disease worse.

Now try this

When are soot and carbon monoxide produced, and are they always formed together?

1 Explain why soot and carbon monoxide are produced during incomplete combustion of a hydrocarbon fuel. **(3 marks)**

2 Explain why carbon monoxide is difficult for our bodies to detect. **(2 marks)**

3 A householder sees soot marks around his gas boiler. He thinks that this shows that carbon monoxide is being produced.
Comment on the accuracy of this thought. **(2 marks)**

Acid rain

Rainwater is naturally acidic, but **acid rain** is more acidic than normal.

Sulfur dioxide

Hydrocarbon fuels may contain impurities such as sulfur compounds. When the fuel burns, the sulfur in these impurities is oxidised to form sulfur dioxide:

$$\text{sulfur} + \text{oxygen} \rightarrow \text{sulfur dioxide}$$
$$S(s) + O_2(g) \rightarrow SO_2(g)$$

Sulfur dioxide is a non-metal oxide:
* It dissolves in rainwater to form an acidic solution.

Reducing environmental damage

The problems caused by acid rain can be reduced in several ways, including:

☑ removing sulfur from petrol, diesel oil and fuel oil at the oil refinery before selling it

☑ preventing sulfur dioxide leaving power station chimneys – 'flue gas desulfurisation'

☑ adding calcium carbonate or calcium hydroxide to fields and lakes to neutralise excess acid from acid rain.

The effects of acid rain

waste gases from power stations and vehicles contain sulfur dioxide

sulfur dioxide dissolves in water in the air

rain is more acidic than normal

acid rain speeds up the weathering of buildings and statues

trees are damaged

rivers, lakes and soils are more acidic, which harms organisms living in them

Worked example

Oxides of nitrogen, NO_x, are atmospheric pollutants. They may contribute to acid rain.

(a) Explain why oxides of nitrogen are produced when hydrocarbon fuels are used in vehicle and aircraft engines. **(2 marks)**

Air goes into the engine so that the fuel can burn. Nitrogen and oxygen from the air react together at the high temperatures in the engine to produce oxides of nitrogen.

(b) Balance the equation below to represent the production of nitrogen dioxide. **(1 mark)**

$$N_2(g) + \underline{2}\,O_2(g) \rightarrow \underline{2}\,NO_2(g)$$

Unlike sulfur dioxide, oxides of nitrogen are **not** produced from impurities in the hydrocarbon fuel.

NO_2 is a non-metal oxide that dissolves in rainwater to form an acidic solution.

Nitrogen dioxide is a toxic, orange-brown gas. Unlike carbon monoxide (produced during incomplete combustion), nitrogen dioxide has a sharp smell. You can revise incomplete combustion on page 87.

Now try this

1 State **two** environmental problems caused by acid rain. **(2 marks)**

Look at the diagram to help you.

2 Explain why, when hydrocarbon fuels are used:
(a) sulfur dioxide is produced **(2 marks)**
(b) oxides of nitrogen are produced. **(2 marks)**

Choosing fuels

Most cars use petrol or diesel oil, but hydrogen may also be used as a fuel for cars.

Fossil fuels

These **fossil fuels** are obtained from crude oil:
- petrol (for cars)
- diesel oil (for some cars and trains)
- fuel oil (for large ships and some power stations).

This fossil fuel is obtained from natural gas:
- methane (for domestic cooking and heating).

Non-renewable resources

Non-renewable resources are used up faster than they are formed.

Crude oil and natural gas take millions of years to form (see page 84 for more about this). The fossil fuels obtained from these resources are being used up very quickly, so they are non-renewable fuels.

Hydrogen

The combustion of hydrogen produces only water vapour:

hydrogen + oxygen → water

$$2H_2(g) + O_2(g) \rightarrow 2H_2O(g)$$

Hydrogen is manufactured in several ways, including:
- electrolysis of water (the reverse of the process above)
- cracking of oil fractions (for more about cracking, see page 90)
- reaction of natural gas with steam:

$$CH_4(g) + 2H_2O(g) \rightarrow CO_2(g) + 4H_2(g)$$

Worked example

Describe three features of a good fuel. **(3 marks)**

A good fuel should:
- burn easily – it should be easy to ignite and stay alight
- not produce soot, smoke or ash
- release a lot of energy when it burns.

A good fuel should also be easy to store and transport safely.

Petrol versus hydrogen

Petrol	Hydrogen
👍 burns easily	👍 burns easily
👍 does not produce ash	👍 does not produce ash or smoke
👎 produces carbon dioxide and carbon monoxide as well as water when it burns	👍 only produces water when it burns
👍 releases more energy per kg when it burns than fuels such as coal or wood	👍 releases nearly three times as much energy per kg as petrol
👍 is a liquid, so it is easy to store and transport	👎 is a gas, so it has to be stored at high pressure
	👎 filling stations would need to be adapted for hydrogen to be used in cars

Now try this

1 Explain why the combustion of petrol produces carbon dioxide, but the combustion of hydrogen does not. **(3 marks)**

2 (a) Suggest reasons to explain why hydrogen produced from crude oil or natural gas may not be renewable. **(2 marks)**

Look at the ways hydrogen can be manufactured.

(b) Carbon dioxide is a greenhouse gas linked to global warming and climate change. Hydrogen is often regarded as a 'carbon-neutral fuel', a fuel with production, transport and use that have no overall emission of carbon dioxide.

Suggest reasons to explain why hydrogen may not actually be carbon neutral. **(2 marks)**

Cracking

Cracking is a process carried out on fractions in oil refineries after fractional distillation.

Making fractions more useful

Cracking involves breaking down larger alkanes into smaller, more useful alkanes and **alkenes**.

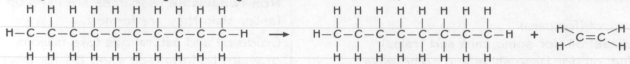

The long molecules are not very useful.

Cracking breaks down molecules by heating them.

Cracking produces shorter-chain alkanes, which are useful as fuels.

Cracking also produces **alkenes**, which are used to make **polymers**.

Cracking in the lab

Paraffin is an alkane. Liquid paraffin can be cracked in the laboratory using this apparatus.

✓ The porous pot catalyst is heated strongly.

✓ The liquid paraffin is heated and evaporates.

✓ The paraffin vapour passes over the hot porous pot and the hydrocarbon molecules break down.

✓ One of the products is ethene, which is a gas and collects in the other tube.

liquid paraffin on mineral wool — broken porous pot — delivery tube — ethene gas — water — HEAT — HEAT

Worked example

The bar chart shows the supply and demand of different fractions from crude oil.

Explain how cracking helps to balance supply with demand. **(2 marks)**

Some larger alkanes, such as bitumen, are in greater supply than their demand. Cracking converts these alkanes into smaller hydrocarbons, such as petrol, which are in greater demand than can be supplied by fractional distillation alone.

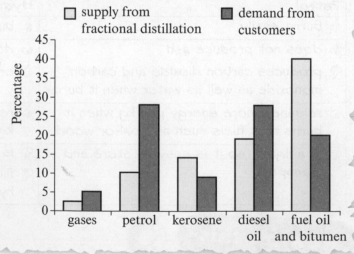

□ supply from fractional distillation   ■ demand from customers

Now try this

1 (a) Give the formula of the substance needed to balance this equation (which represents a cracking reaction):

$C_8H_{18} \rightarrow C_6H_{14} +$ .... **(1 mark)**

(b) State a use for the hydrocarbon in your answer to (a), other than as a fuel. **(1 mark)**

Study the diagram at the top of the page.

Extended response – Fuels

There will be one or more 6-mark question on your exam paper. For these questions, you will need to think scientifically and structure your answer logically, showing how the points you make are related to each other.

You can revise the topics for this question, which is about **fuels**, on pages 84–89.

Worked example

Crude oil is a complex mixture of hydrocarbons. Most belong to the alkane homologous series. Diesel oil is a fuel produced from crude oil.

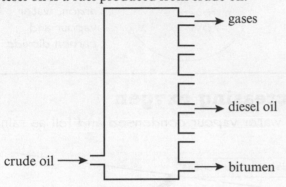

Explain how diesel oil is separated from crude oil. Use your knowledge and understanding, and the diagram above, in your answer. **(6 marks)**

Crude oil is separated into different fractions by fractionation distillation. The crude oil is heated to around 350 °C to evaporate it. Some of the hydrocarbons stay in the liquid state and leave at the bottom of the column. The other hydrocarbons rise inside the fractionating column in the gas state.

The fractionating column has a temperature gradient inside it. The column becomes cooler towards the top, so the hydrocarbons cool down as they rise. They condense to the liquid state when they reach a part of the column that is cool enough.

The gases do not condense and they leave at the top. Diesel oil leaves through a pipe as a liquid fraction when it condenses.

Command word: Explain

When you are asked to **explain** something, it is not enough just to state or describe it.

Your answer must contain some reasoning or justification of the points you make.

The labelled diagram in the stem of the question helps you in two ways:
- It may remind you that crude oil is separated by fractionation distillation (not, for example, by cracking).
- It shows approximately where diesel oil leaves the fractionating column.

You should state clearly which separation method is involved, and outline what happens to the crude oil immediately after heating it.

The temperature gradient inside a fractionating column is important to the process. You could also mention that the hydrocarbons condense when they cool to their boiling point, because this is an important feature too.

Your answer could also explain that smaller hydrocarbons have lower boiling points because they have weaker intermolecular forces.

Your answer should show comprehensive knowledge and understanding of the topic covered. Take care to use the information given to you in support of your explanations. Organise your answer in a structured way with clear lines of reasoning.

Now try this

Diesel oil is used as a fuel for some cars and trains. Scientists and engineers are researching ways to use hydrogen as a fuel for cars instead of diesel oil. Evaluate the use of hydrogen as a fuel for cars, rather than diesel oil. **(6 marks)**

You must be able to evaluate the use of hydrogen rather than petrol as a fuel for cars. Therefore, here you can discuss its advantages and disadvantages compared with a different fossil fuel.

The early atmosphere

Gases produced by volcanic activity formed the Earth's early atmosphere.

Gases in the early atmosphere

Scientists believe that the Earth's early atmosphere, billions of years ago, contained:
- little or no oxygen
- a large amount of carbon dioxide
- water vapour
- small amounts of other gases.

Evidence for this includes:
- the mixture of gases released by volcanoes
- the atmospheres of other planets in our solar system today, which have not been changed by living organisms.

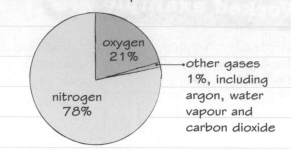

Today's atmosphere

Nitrogen and oxygen are the main gases in the modern atmosphere.

oxygen 21%
nitrogen 78%
other gases 1%, including argon, water vapour and carbon dioxide

Decreasing carbon dioxide and increasing oxygen

The Earth was very hot to start with. As it cooled, water vapour condensed and fell as rain to form the oceans.

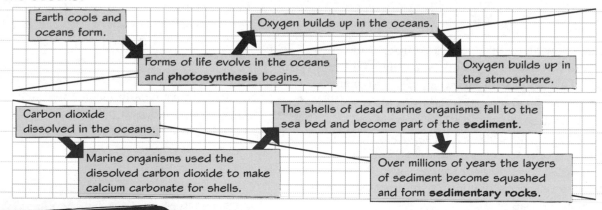

Earth cools and oceans form.

Forms of life evolve in the oceans and **photosynthesis** begins.

Oxygen builds up in the oceans.

Oxygen builds up in the atmosphere.

Carbon dioxide dissolved in the oceans.

Marine organisms used the dissolved carbon dioxide to make calcium carbonate for shells.

The shells of dead marine organisms fall to the sea bed and become part of the **sediment**.

Over millions of years the layers of sediment become squashed and form **sedimentary rocks**.

Worked example

Some rocks contain iron. Very old rocks do not contain iron oxides but later ones do.

(a) Explain how these rocks provide evidence for changes to the Earth's atmosphere. **(3 marks)**

The very old rocks are evidence that the early atmosphere contained little or no oxygen, because they do not contain iron oxides. Later rocks do contain iron oxides, which is evidence that oxygen was released into the atmosphere and reacted with iron in the rocks.

(b) Describe a simple test for oxygen. **(2 marks)**

A glowing wooden splint relights in oxygen.

When you describe a laboratory test, say what you would do, and what you would observe.

Remember:
- a lighted wooden splint ignites hydrogen with a pop
- limewater turns milky in the presence of carbon dioxide.

When oxygen was first produced, it reacted with iron in rocks to produce iron oxides. This meant that, even though primitive plants were photosynthesising, oxygen levels in the atmosphere did not begin to rise straight away.

Now try this

1 (a) State how the formation of oceans was a cause of decreasing carbon dioxide levels. **(1 mark)**

(b) Describe why the levels of oxygen in the atmosphere increased. **(2 marks)**

2 Suggest a reason that explains why scientists cannot be certain about the Earth's early atmosphere. **(1 mark)**

Greenhouse effect

The greenhouse effect helps to keep the Earth warm enough for living organisms to exist.

Reducing radiation to space

In the **greenhouse effect**:
- **Greenhouse gases** in the atmosphere absorb heat radiated from the Earth.
- The greenhouse gases then release energy in all directions.

This reduces the amount of heat radiated into space, keeping the Earth warm.

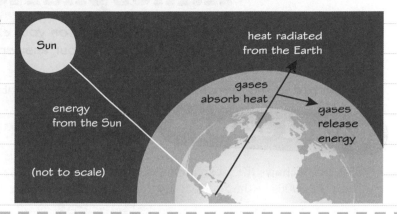

Sun

heat radiated from the Earth

energy from the Sun

gases absorb heat

gases release energy

(not to scale)

Greenhouse gases

The table shows gases that are particularly good at absorbing and emitting energy by radiation.

Greenhouse gas	Typical source
carbon dioxide	burning fossil fuels
methane	livestock farming, e.g. cattle
water vapour	evaporation from oceans

This is not seen as a problem because excess water vapour leaves the atmosphere as rain and snow.

Global warming and climate change

The accumulation of greenhouse gases in the atmosphere:
- increases the greenhouse effect.

This:
- increases the warming effect in the atmosphere – **global warming**.

Global warming is associated with:
- long-term changes to weather patterns – **climate change**
- rising sea levels due to melting ice and expanding ocean water.

Worked example

Evaluate how far the graph provides evidence for global warming.   **(2 marks)**

The graph shows that both temperature and carbon dioxide concentration have been rising since 1950. There is a correlation but this does not <u>prove</u> that rising carbon dioxide levels are causing the warming.

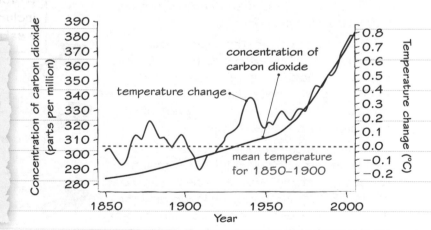

Concentration of carbon dioxide (parts per million)

Temperature change (°C)

concentration of carbon dioxide

temperature change

mean temperature for 1850–1900

Year

Many scientists believe that carbon dioxide is the main cause of global warming, but this opinion is based on computer modelling and other information, not just on graphs like this.

Now try this

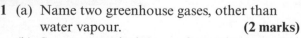

1 (a) Name two greenhouse gases, other than water vapour.   **(2 marks)**
  (b) State one typical source for each gas named in your answer to **(a)**.   **(2 marks)**

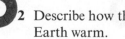

2 Describe how the greenhouse effect keeps the Earth warm.   **(2 marks)**
3 Explain how the increased use of fossil fuels, which contain carbon and carbon compounds, may lead to global warming.   **(4 marks)**

Extended response – Atmospheric science

There will be one or more 6-mark question on your exam paper. For these questions, you will need to think scientifically and structure your answer logically, showing how the points you make are related to each other.

You can revise the topics for this question, which is about **the atmosphere**, on pages 92 and 93.

Worked example

The graph shows the mean annual release of carbon dioxide from burning fossil fuels during each decade from the 1900s to the 2000s. It also shows the mean change in global temperature during that time.

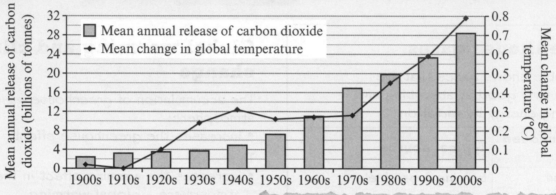

Evaluate this evidence of human activity as a cause of global warming. **(6 marks)**

The mass of carbon dioxide released increased each decade from the 1900s to the 2000s. For example, in the 1900s about 2.5 billion tonnes was released on average each year, but by the 2000s it was over 28 billion tonnes, more than 10 times greater.

Over the same period, the mean global temperature increased by nearly 0.8 °C. As the carbon dioxide was released by burning fossil fuels, this shows that there is a positive correlation between the consumption of fossil fuels and temperature change.

However, the temperature increased more rapidly in the 1930s and 1940s than the release of carbon dioxide. It also went down in the 1950s and barely changed until the 1980s, whereas the release of carbon dioxide increased greatly. By itself, the evidence given does not show that human activity caused the global warming observed, but it strongly suggests that it did.

Command word: Evaluate

When you are asked to **evaluate** something, you need to review information given to you, and then bring it together to form a conclusion. You should include evidence such as strengths and weaknesses, and arrive at a judgement.

You should use the information in your answer, e.g. by interpreting the bar chart.

A **correlation** between two sets of data means that they are connected in some way. In this case, there is a **positive correlation** – as one variable increases, the other one does too.

You should point out evidence for weaknesses as well as strengths for the main argument.

A **causal** relationship means that a change in one variable causes a change in another variable.

Now try this

The concentration of carbon dioxide in the atmosphere increased by about 35% between 1900 and 2015.

Describe the processes that remove carbon dioxide from the atmosphere, and suggest reasons to explain the observed increase in its concentration. **(6 marks)**

Tests for metal ions

Practical skills Metal ions can be identified using simple laboratory tests.

Flame tests

Some metal ions can be identified using **flame tests**. Different metal ions in compounds produce different and distinctive colours in flame tests.

Metal ion	Flame test colour
lithium, Li^+	red
sodium, Na^+	yellow
potassium, K^+	lilac
calcium, Ca^{2+}	orange–red
copper, Cu^{2+}	blue–green

Clean the flame test loop in acid each time, rinse with water and check it is clean in the Bunsen burner flame.

To test a substance, dip the clean loop in a solution of the ions and hold at the edge of a blue flame.

Hydroxide precipitates

Some metal ions form coloured hydroxide **precipitates**. The sample solution is placed in a test tube and a few drops of dilute sodium hydroxide solution are added. The table shows the colours you need to know.

Some metal ions form white precipitates with sodium hydroxide solution:
- calcium ions, Ca^{2+}
- aluminium ions, Al^{3+}.

Metal ion	Colour of precipitate
copper, Cu^{2+}	blue
iron(II), Fe^{2+}	green
iron(III), Fe^{3+}	brown

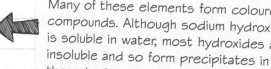

Copper and iron are **transition** metals. Many of these elements form coloured compounds. Although sodium hydroxide is soluble in water, most hydroxides are insoluble and so form precipitates in these tests.

Worked example

Describe how to distinguish between the hydroxide precipitate formed by aluminium ions and that formed by calcium ions. **(2 marks)**

Add excess sodium hydroxide solution. The aluminium hydroxide precipitate dissolves but the calcium hydroxide precipitate does not.

Aluminium and calcium are **not** transition metals, so they form white or colourless compounds. Both these precipitates are white.

The aluminium hydroxide precipitate reacts with the sodium hydroxide, forming a colourless solution.

Now try this

 1 Describe a laboratory test to distinguish between iron(II) chloride solution and iron(III) chloride solution. **(3 marks)**

2 Suggest a reason that explains why it is difficult to identify a mixture of ions in a flame test. **(1 mark)**

 3 Describe **two** different ways in which copper ions, Cu^{2+}, may be detected. **(2 marks)**

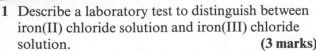

State what you would do and what you expect to see.

More tests for ions

 Practical skills Non-metal ions can be identified using simple laboratory tests.

Tests for sulfate and carbonate ions

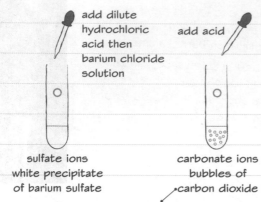

add dilute hydrochloric acid then barium chloride solution

add acid

sulfate ions
white precipitate of barium sulfate

carbonate ions
bubbles of carbon dioxide

Carbon dioxide turns limewater milky.

Tests for halide ions

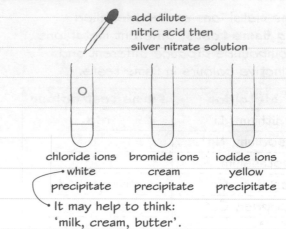

add dilute nitric acid then silver nitrate solution

chloride ions
white precipitate

bromide ions
cream precipitate

iodide ions
yellow precipitate

It may help to think: 'milk, cream, butter'.

Test for ammonium ions

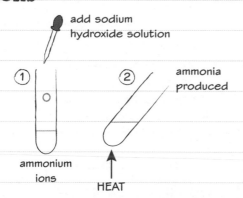

add sodium hydroxide solution

① ②

ammonia produced

ammonium ions

HEAT

Test for ammonia gas

In the test for ammonium ions, the reaction produces water and ammonia:

$$NH_4^+(aq) + OH^-(aq) \rightarrow H_2O(l) + NH_3(g)$$

You can detect the ammonia gas produced:

✓ Damp red litmus paper turns blue.

✓ Hydrogen chloride gas (from concentrated hydrochloric acid) reacts with ammonia to form a white smoke of ammonium chloride:

$$NH_3(g) + HCl(g) \rightarrow NH_4Cl(s)$$

Ammonia has a characteristic sharp smell.

Worked example

Silver nitrate solution is used to test for the presence of chloride ions. A white precipitate of silver chloride forms:

$$Ag^+(aq) + Cl^-(aq) \rightarrow AgCl(s)$$

Carbonate ions also form a precipitate:

$$2Ag^+(aq) + CO_3^{2-}(aq) \rightarrow Ag_2CO_3(s)$$

Explain why, when carrying out a test for chloride ions, dilute nitric acid is added to the sample before adding silver nitrate solution. **(2 marks)**

The nitric acid reacts with the carbonate ions. As the sample no longer contains carbonate ions, they cannot form a precipitate, which gives a false-positive result.

You cannot acidify the sample with dilute hydrochloric acid because this contains chloride ions. These ions would give a false-positive result. It is important that the test for an ion is unique, and detects only the intended ion.

Now try this

Explain how you could carry out a test to distinguish between sodium chloride solution and sodium iodide solution.
(3 marks)

Instrumental methods

Instrumental methods of analysis use machines to detect and analyse substances.

Advantages of instrumental methods

Compared with simple chemical tests and analysis, instrumental methods offer improved:

 sensitivity – they can detect and analyse very small amounts of different substances

2 **accuracy** – they measure amounts of different substances very accurately

3 **speed** of tests – they carry out each analysis quickly, and the machines can run all the time.

Some examples

The table gives some examples of instrumental methods of analysis. You do not need to recall this information.

Name of method	Information given
mass spectrometry	measures M_r values and identifies compounds
infrared spectroscopy	detects covalent bonds and identifies compounds
gas chromatography	measures amounts of each component in mixtures

Flame photometer

Flame photometry is an instrumental method of analysis based on flame tests (look at page 95 for more about flame tests):
• The sample is vaporised in a hot flame.
• A **spectrum** of the light emitted by metal ions is produced.
• The brightness of a particular wavelength is measured.

The data from a flame photometer can:

1 **identify** the metal ions present in the sample by comparing the spectrum with a **reference spectrum** from a known substance

2 **determine** the concentration of ions in a solution using a **calibration curve**.

Note that you do not need to be able to recall how the flame photometer works.

look at page 95

Worked example

The emission spectra on the right were produced by flame photometry. The **mix** contains two different metal ions. Identify these ions. **(2 marks)**

Li⁺ and K⁺.

The spectrum produced by the mixture has the two lines from Li⁺, and the lines from K⁺. It does not show the lines produced by the other ions (Na⁺, Ca²⁺ or Cu²⁺).

If the lines in a spectrum match the lines in a reference spectrum, the ions must be the same.

Now try this

1 Solutions containing different concentrations of Na⁺ ions were analysed by flame photometry.

The readings were plotted against the concentration of Na⁺ ions to produce this calibration curve.

Another solution containing Na⁺ ions gave a reading of 60%.

Use the graph to determine the concentration of Na⁺ ions in this solution. **(1 mark)**

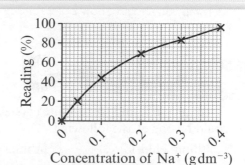

Extended response – Tests for ions

There will be one or more 6-mark question on your exam paper. For these questions, you will need to think scientifically and structure your answer logically, showing how the points you make are related to each other.

You can revise the topics for this question, which is about **qualitative analysis: tests for ions**, on pages 95 and 96.

Worked example

The labels have fallen off four different containers of white powders:

ammonium chloride NH₄Cl	sodium chloride NaCl

sodium sulfate Na₂SO₄	sodium carbonate Na₂CO₃

Explain how, using chemical tests, you could identify which container held which substance.

(6 marks)

First carry out a flame test on the four solids. The three sodium compounds will produce a yellow flame due to their sodium ions. The fourth compound should be ammonium chloride.

The identity of the ammonium chloride can be confirmed by adding sodium hydroxide solution. When the mixture is warmed, ammonia should be given off, turning damp red litmus paper blue.

Dissolve the three sodium compounds in water, and then add dilute hydrochloric acid. The sodium carbonate solution will produce bubbles of a gas (carbon dioxide) that turns limewater milky. Add barium chloride solution to the remaining solutions. The sodium sulfate solution will produce a white precipitate.

The last solution should be sodium chloride solution. To check that it is sodium chloride, dissolve another sample of the powder in water. Add dilute nitric acid, then silver nitrate solution. It should produce a white precipitate of silver chloride.

Command word: Explain

When you are asked to **explain** something, it is not enough just to state or describe it. Your answer must contain some reasoning or justification of the points you make.

The labels help you to work out which ions are present. You can start to plan a suitable series of tests to distinguish between them. There may be more than one way to use the tests you know about in these questions.

The flame test would work on the substances in the solid state or in solution. All four are soluble:
- all ammonium and sodium salts are soluble (you can revise solubility rules on page 40).

A negative result does not really identify an ion. You should plan a **confirmatory test** to obtain a positive result for that ion.

You could use dilute nitric acid. Then, carry out a silver nitrate test for chloride ions instead at this point.

You need to plan for another sample to be dissolved for testing: hydrochloric acid (added in the previous step) contains chloride ions and would give a false-positive result.

Your answer should show comprehensive knowledge and understanding of the topic covered. It is important in this sort of question to organise your answer in a structured way, giving clear lines of reasoning.

Now try this

A technician found some white crystals in an unlabelled beaker in a laboratory. He knew that the substance could be one of the following:
- lithium chloride
- lithium carbonate
- potassium chloride
- potassium iodide.

Devise a series of tests that the technician could carry out to identify the substance. **(6 marks)**

More about alkanes

The **alkanes** are a homologous series of saturated hydrocarbons.

Molecular formulae

The **general formula** for the alkanes is:

• C_nH_{2n+2}

where n = number of carbon atoms.

You can use this general formula to predict the **molecular formula** of an alkane. Butane molecules have four carbon atoms, so:

• $n = 4$
• $2n + 2 = (2 \times 4) + 2 = 10$
• molecular formula is C_4H_{10}.

You can revise other features of alkanes and homologous series on page 86.

Drawn structures

To draw the structure of a butane molecule:

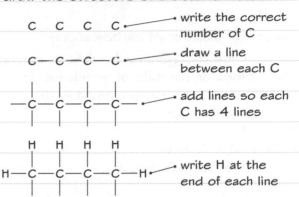

The first three alkanes

	Name	Molecular formula	Drawn structure	
meth = 1 carbon atom	methane	CH_4	H—C—H (with H above and below)	methane has no C–C bonds
eth = 2 carbon atoms	ethane	C_2H_6	H—C—C—H (with H's)	saturated (no C=C bonds, just C–C bonds)
prop = 3 carbon atoms	propane	C_3H_8	H—C—C—C—H (with H's)	each successive alkane differs by CH_2

Worked example

Write a balanced equation for the complete combustion of butane.   **(2 marks)**

$$C_4H_{10} + 6\tfrac{1}{2}O_2 \rightarrow 4CO_2 + 5H_2O$$

If you prefer, multiply through by 2 to get:

$$2C_4H_{10} + 13O_2 \rightarrow 8CO_2 + 10H_2O$$

When you balance equations for the complete combustion of a hydrocarbon:

1 number of CO_2 = number of C atoms

2 number of H_2O = half the number of H atoms

3 number of O_2 = half the total of O atoms on the right

Now try this

1 Hexane is an alkane with six carbon atoms.
   (a) Give the molecular formula of hexane, and draw its structure.   **(2 marks)**

   (b) Explain why hexane is 'saturated'.   **(1 mark)**
   (c) Write a balanced equation for the complete combustion of hexane.   **(2 marks)**

Alkenes

The **alkenes** are a homologous series of unsaturated hydrocarbons.

Molecular formulae

The **general formula** for the alkenes is:
- C_nH_{2n}

where n = number of carbon atoms.

You can use this general formula to predict the **molecular formula** of an alkene. For example, propene molecules have three carbon atoms, so:
- $n = 3$
- $2n = 2 \times 3 = 6$
- molecular formula is C_3H_6.

Drawn structures

To draw the structure of a propene molecule:

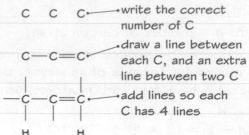

- write the correct number of C
- draw a line between each C, and an extra line between two C
- add lines so each C has 4 lines
- write H at the end of each line

Reactions of alkenes

In the **complete combustion** of alkenes:
- carbon is oxidised to carbon dioxide
- hydrogen is oxidised to water vapour.

For example:

ethene + oxygen → carbon dioxide + water

$$C_2H_4 + 3O_2 \rightarrow 2CO_2 + 2H_2O$$

Alkenes react with **bromine** to produce colourless compounds, e.g.

ethene + bromine → dibromoethane

$$C_2H_4 + Br_2 \rightarrow C_2H_4Br_2$$

Test for alkenes

Alkenes react with bromine because they are **unsaturated** (their molecules contain the functional group, C=C):

This is the basis of a test to distinguish between alkanes and alkenes:

- ✓ Add a few drops of bromine water.
- ✓ It stays orange in an alkane.
- ✓ It is decolourised in an alkene.

Worked example

 (a) Write the molecular formula of butene. **(1 mark)**

C_4H_8

 (b) The functional group in a butene molecule can be in two different places. Draw the structures of the two forms of butene, showing all their covalent bonds. **(2 marks)**

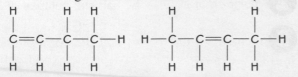

The 'but' in the name means four carbon atoms. The number of hydrogen atoms in alkene molecules is twice the number of carbon atoms.

The question is describing the two **isomers** of butene – compounds with the same molecular formula but a different arrangement of atoms.

Their names include a number to identify the position of the C=C bond. From left to right:
- but-1-ene and but-2-ene.

Now try this

1 Hexane, C_6H_{14}, and hexene, C_6H_{12}, are both liquids at room temperature.
   (a) Explain why hexene is described as an unsaturated hydrocarbon. **(3 marks)**
   (b) Describe a chemical test to distinguish between hexane and hexene. **(3 marks)**

Addition polymers

Addition polymers are relatively large molecules made by combining smaller molecules containing C=C bonds.

Polymerisation

A **polymer** is a substance:
- of high average relative molecular mass, M_r
- made up of small **repeating units**.

Poly(ethene) is a polymer made by combining many ethene molecules (**monomers**).

monomers → polymer

Equations for polymerisation

You show equations for addition polymerisation using structures like this:

n = a large number.

→ This is a repeating unit.

The equation models many (n) ethene monomers combining to form a poly(ethene) molecule containing n repeating units.

Worked example

The diagram shows the structure of the repeating unit of poly(chloroethene).

Draw the structure of the monomer. **(1 mark)**

> Poly(chloroethene) is also called PVC (after its older name, polyvinyl chloride).

> To deduce the structure of a monomer from the structure of a polymer, draw:
> - the symbols for the atoms in the same positions
> - all the single bonds except for the two that pass through the large brackets.
>
> Then change the C–C bond to a C=C bond.

Properties and uses

Different polymers have different properties, so they have different uses.

Polymer	Properties	Uses
poly(ethene)	flexible, cheap, good electrical insulator	plastic bags, plastic bottles and Clingfilm
poly(propene)	flexible, shatterproof, has a high softening point	buckets and bowls
poly(chloroethene) (PVC)	tough, cheap, long lasting, good electrical insulator	window frames, gutters and pipes, insulation for electrical wires
PTFE poly(tetrafluoroethene)	tough, slippery, resistant to corrosion, good electrical insulator	non-stick coatings for frying pans, containers for corrosive substances, insulation for electric wires

---

Now try this

1 The diagram shows the structure of propene.
  Write an equation to show the formation of poly(propene). **(2 marks)**

Condensation polymers

Polyesters are **condensation polymers** rather than addition polymers.

Two monomers

Polyesters need two different monomers:

 a molecule containing two **carboxylic acid** groups:

differs between monomers   carboxylic acid group

 a molecule containing two **alcohol** groups:

differs between monomers   alcohol group

Forming an ester link

An **ester link** forms each time the two different monomers react together:

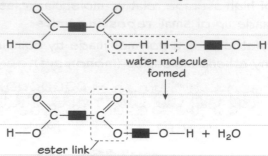

water molecule formed

ester link

✓ One molecule of water forms each time an ester link forms.

Condensation polymerisation

When the two different monomers react together, producing a water molecule:

- the other molecule has a carboxylic acid group and an alcohol group
- these groups can react with more monomers
- this process continues, producing a very long **polyester** molecule.

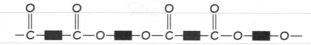

Worked example

Complete the table to show the differences between addition polymerisation and condensation polymerisation. **(2 marks)**

	Addition	Condensation
Number of different monomers	one	two
Number of different products	one	two

The monomers for each type are:
- addition polymerisation – alkenes
- condensation polymerisation – 'dicarboxylic acids' and 'diols'.

The products for each type are:
- addition polymerisation – polymer only
- condensation polymerisation – polymer and water.

Now try this

1 PET is a polyester used to make drinks bottles and fleece linings.

(a) State whether PET is an addition polymer or a condensation polymer. **(1 mark)**

(b) State the type of monomers needed to manufacture PET. **(2 marks)**

(c) Name the product formed, other than PET, during the manufacture of PET. **(1 mark)**

Biological polymers

Biological polymers are naturally occurring condensation polymers.

DNA

DNA (deoxyribonucleic acid) is found in the nucleus of cells. It has a 'double helix' structure.

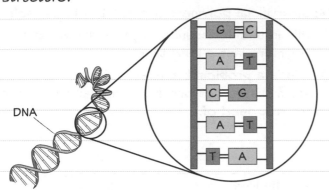

DNA

DNA is a polymer made from monomers called **nucleotides**.

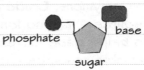

phosphate   base   sugar

There are four different nucleotides that differ in the 'base' they contain.

You do not need to know the names of the bases but they are usually abbreviated to A, T, G and C.

Proteins

Proteins are polymers made from monomers called **amino acids**.

amino acids

Each amino acid molecule (as in alanine below) has two reactive functional groups, which allow many amino acids to bond together to form a protein.

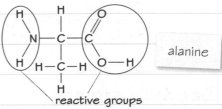

alanine

reactive groups

Worked example

Starch is a substance found in plant cells.

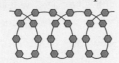

Describe the structure of starch.   **(2 marks)**

Starch is a polymer made from many sugar monomers joined together.

The sugar monomer in starch is **glucose**. The carbon atoms in glucose, $C_6H_{12}O_6$, are arranged in a ring, so each glucose molecule is modelled as a hexagon in the diagram.

There may be thousands of these monomers in a single starch molecule.

Now try this

1 Draw one straight line from the name of each polymer to its monomer.   **(3 marks)**

monomer	polymer
sugars	protein
alkenes	DNA
amino acids	starch
nucleotides	

Polymer problems

There are problems associated with the manufacture and use of polymers.

Manufacture

Crude oil is the main raw material needed to make addition polymers and most condensation polymers. Crude oil is:

- a finite resource – it is no longer being made or is being made extremely slowly
- a non-renewable resource – it is being used up faster than it is formed
- often imported by users – its supply and cost vary over time.

Non-biodegradability

Biodegradable materials eventually rot away:

- microbes feed on them
- this breaks them down.

Most artificial polymers are not biodegradable:

- 👍 This is useful because objects made from polymers last a long time.
- 👎 This is a problem because objects made from polymers do not break down easily when they are disposed of.

Disposing of polymers

Landfill sites 👎
- polymers are not biodegradable
- they last for many years
- we are running out of landfill sites

Burning 👎
- many polymers release toxic gases when they burn

Disposing of polymers

Recycling 👍
- melting and reforming into new objects
- breaking down into new raw materials

Biodegradable polymers 👍
- these are being developed
- they will rot away in landfill sites

Landfill sites

Most waste goes into **landfill**:

- 👍 Waste is disposed of quickly.
- 👍 Waste is out of sight once it is covered over.
- 👎 Space for landfill sites is running out.
- 👎 Most polymers are not biodegradable and will last for many years.
- 👎 Landfill sites are unsightly and attract pests.

Worked example

Give **two** advantages and **two** disadvantages of the disposal of polymers by recycling rather than by landfill.   **(4 marks)**

Dumping polymers in landfill sites is a waste of a non-renewable resource, as polymers are made from crude oil.

Recycling means that less waste goes into landfill. On the other hand, recycling is expensive because the different polymers must be collected and sorted.

Some polymers cannot be recycled.

The use of the connective 'On the other hand' is a useful way to show where the answer changes from advantages to disadvantages.

Now try this

1. Describe the advantages and disadvantages of disposing of waste polymers by:
   (a) combustion (burning them in incinerators)   **(2 marks)**
   (b) burying them in landfill sites.   **(2 marks)**

2. State **two** ways in which waste polymers may be made into new objects.   **(2 marks)**

3. Suggest reasons that explain why it has become worthwhile for companies in China to pay for waste poly(ethene) to be shipped to them from America and Europe.   **(2 marks)**

Extended response – Hydrocarbons and polymers

There will be one or more 6-mark question on your exam paper. For these questions, you will need to think scientifically and structure your answer logically, showing how the points you make are related to each other.

You can revise the topics for this question, which is about **hydrocarbons** and **polymers**, on pages 99–101.

Worked example

Tetrafluoroethene, C_2F_4, is used to make poly(tetrafluoroethene). This is an addition polymer often called PTFE. It is used as the non-stick coating in pans and for lining pipes carrying corrosive liquids.

Explain how PTFE molecules are formed from tetrafluoroethene molecules, and how the properties of PTFE are related to its uses.

State **one** problem associated with the disposal of PTFE, and give a reason for your answer.

(6 marks)

Tetrafluoroethene can act as a monomer because its molecules contain a C=C bond. This bond can open up, allowing many tetrafluoroethene molecules to react together to form a long PTFE polymer molecule:

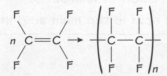

PTFE is very slippery, so food will not stick to it in a pan. Its molecules consist of long chains of carbon atoms, with strong intermolecular forces between molecules, so its melting point is high. PTFE does not react with acids and alkalis, making it suitable for lining chemical pipes.

PTFE is not biodegradable. This means that it will persist in the environment if disposed of in a landfill site. It may produce harmful gases if it is disposed of by combustion in an incinerator.

Command word: State

If you are asked to **state** something, you need to recall one or more pieces of information. The command words **give** and **name** are similar to **state**.

Command word: Give a reason

When you have given an answer, you need to say why you gave this answer or why this answer is correct.

You should explain why tetrafluoroethene molecules act as monomers. You could also state that the reaction is addition polymerisation.

Remember that balanced equations help you to model reactions. In this example, the equation includes the repeating unit for PTFE. You can work out the structure of tetrafluoroethene from the formula and name given to you.

You could point out that the high melting point means that the PTFE coating will stay solid during cooking, rather than melting and leaving the pan. Its unreactivity is another reason why it is suitable for use with food.

The question asks you to state one problem but two are given instead.

Your answer should show comprehensive knowledge and understanding of the topic covered, organised in a structured way with clear lines of reasoning. Take care not to give more answers than asked for.

Now try this

Both compounds are hydrocarbons with two carbon atoms in their molecules.

Ethane and ethene are compounds produced from crude oil. Both undergo complete combustion but only one of them can take part in an addition reaction with bromine.

Describe these reactions, and explain how to distinguish between the two compounds using a simple laboratory test. In your answer, draw their structures and include balanced equations. **(6 marks)**

Alcohols

The alcohols form a homologous series of compounds.

Homologous series

The **alcohols**:
- have the same functional group, –OH
- have similar chemical properties
- differ in the molecular formulae of neighbouring members by CH_2
- show a gradual variation in physical properties, such as boiling points.

You can revise other homologous series: alkanes on page 86, alkenes on page 100 and carboxylic acids on page 108.

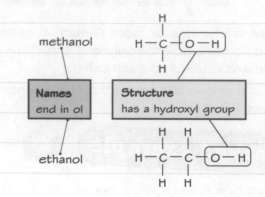

methanol

Names	Structure
end in ol	has a hydroxyl group

ethanol

Reactions of alcohols

Methanol, ethanol and propanol all:
- dissolve in water to form a neutral solution
- react with sodium to produce hydrogen
- burn in air.

For example:

$$methanol + oxygen \rightarrow carbon\ dioxide + water$$

$$CH_3OH(l) + 1\tfrac{1}{2}O_2(g) \rightarrow CO_2(g) + 2H_2O(l)$$

Remember that there is an oxygen atom in the alcohol when balancing these equations.

Oxidation of ethanol

Burning, or combustion, is an example of an oxidation reaction. Ethanol burns in air:

$$ethanol + oxygen \rightarrow carbon\ dioxide + water$$

$$C_2H_5OH(l) + 3O_2(g) \rightarrow 2CO_2(g) + 3H_2O(l)$$

Ethanol can also be oxidised to **ethanoic acid** by:

☑ chemicals called **oxidising agents**, or

☑ the action of microbes.

Ethanoic acid is the main acid in vinegar.

Worked example

(a) Draw the structure of a molecule of propanol, C_3H_7OH. Show all the covalent bonds. **(1 mark)**

```
    H   H   H
    |   |   |
H — C — C — C — O — H
    |   |   |
    H   H   H
```

This compound is more accurately called propan-1-ol. The number 1 in the name tells you that the –OH group is attached to the first carbon atom in the molecule.

You do not need to be able to name or draw propan-2-ol, but its –OH group is attached to the middle carbon atom instead.

(b) Predict the products of the complete combustion of propanol. **(2 marks)**

Carbon dioxide and water vapour.

You should be able to predict the products of different members of the same homologous series.

Now try this

1 The formula for butanol (also called butan-1-ol) can be shown as C_4H_9OH.

(a) Explain how you know that butanol is a member of the alcohol homologous series. **(2 marks)**

(b) Draw the structure of a molecule of butanol. Show all the covalent bonds. **(2 marks)**

(c) Write a balanced equation for the complete combustion of butanol. **(2 marks)**

Making ethanol

Ethanol is the alcohol found in wine, beer and other alcoholic drinks.

Fermentation

Ethanol, C_2H_5OH, is produced from carbohydrates in aqueous solution by a process called **fermentation**. Carbon dioxide is also produced in this reaction. The carbohydrates can be sugars from fruit, such as grapes, or from the breakdown of starch from wheat or barley.

Yeast is a single-celled fungus. It provides enzymes for fermentation to happen:

$$glucose \rightarrow ethanol + carbon\ dioxide$$

$$C_6H_{12}O_6(aq) \rightarrow 2C_2H_5OH(aq) + 2CO_2(g)$$

Worked example

Explain why the fermentation mixture must be kept warm and under anaerobic conditions.

(4 marks)

The reaction is too slow at low temperatures and the yeast enzymes do not work at high temperatures. Fermentation is a type of anaerobic respiration. If oxygen is present, aerobic respiration happens instead, producing only carbon dioxide and water.

Remember that the enzymes in yeast do not work at high temperatures because they are denatured. The optimum temperature for fermentation is around 35 °C.

Fractional distillation

Fractional distillation is used to obtain a concentrated solution of ethanol. This works because:

- ethanol has a lower boiling point than water.

There is more information about fractional distillation and its use on pages 28 and 85.

Note that it is not possible to obtain **pure** ethanol by this method alone. The small amount of water that remains must be absorbed chemically.

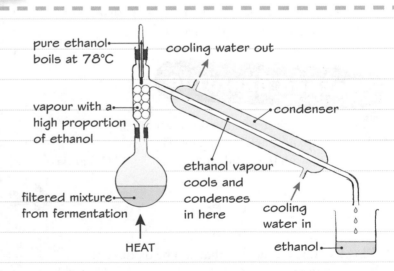

pure ethanol boils at 78°C

cooling water out

vapour with a high proportion of ethanol

condenser

ethanol vapour cools and condenses in here

cooling water in

filtered mixture from fermentation

HEAT

ethanol

Now try this

1 Ethanol is produced by fermentation.
   (a) Write a balanced equation for the process, including state symbols.   **(3 marks)**

   (b) Describe the catalyst used for this process.   **(2 marks)**
   (c) Suggest a suitable temperature for carrying out fermentation.   **(1 mark)**

2 Explain how concentrated ethanol is obtained from the fermentation mixture.   **(2 marks)**

Carboxylic acids

The carboxylic acids form a homologous series of compounds.

Homologous series

The **carboxylic acids**:
- have the same functional group, –COOH
- have similar chemical properties
- differ in the molecular formulae of neighbouring members by CH_2
- show a gradual variation in physical properties, such as boiling points.

You can revise other homologous series on these pages: alkanes page 86, alkenes page 100 and alcohols on page 106.

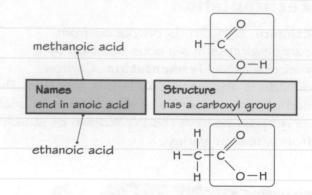

methanoic acid

ethanoic acid

Names	Structure
end in anoic acid	has a carboxyl group

Reactions of carboxylic acids

Carboxylic acids have the typical chemical properties of **acids**. They can:
- react with carbonates to produce a salt, water and carbon dioxide
- react with magnesium and other reactive metals to produce a salt and hydrogen
- dissolve in water to produce acidic solutions (pH less than 7).

You can revise the properties and reactions of acids on pages 34 and 36.

Weak acids

Carboxylic acids are **weak acids**. This is because they only partially dissociate into ions when they dissolve in water, e.g.

ethanoic acid ⇌ ethanoate ion + hydrogen ion

$$CH_3COOH(aq) \rightleftharpoons CH_3COO^-(aq) + H^+(aq)$$

At the same concentration, weak acids have a higher pH (they are less acidic) than strong acids such as hydrochloric acid (see pages 34 and 35).

Worked example

(a) Draw the structure of a molecule of propanoic acid, C_2H_5COOH. Show all the covalent bonds. **(2 marks)**

When you draw the structure, make sure that you show all the covalent bonds in the molecule. Notice how the –COOH group is shown. The bonds present in it are:
- C=O
- C–O
- O–H

(b) Name the alcohol that can be oxidised to produce propanoic acid. **(1 mark)**

propanol

Alcohols can be oxidised to produce carboxylic acids, e.g.

ethanol + oxidising agent → ethanoic acid + water

$$CH_3CH_2OH + 2[O] \rightarrow CH_3COOH + H_2O$$

You do not need to know the oxidising agent used, but you should be able to extend the idea to other alcohols.

Now try this

1 The formula for butanoic acid is C_3H_7COOH.
(a) Explain how you know that butanoic acid is a member of the carboxylic acid homologous series. **(2 marks)**
(b) Draw the structure of a molecule of butanoic acid. Show all the covalent bonds. **(2 marks)**
(c) Name the alcohol that can be oxidised to produce butanoic acid. **(1 mark)**

🧪 Practical skills Investigating combustion

Core practical

Investigating combustion

Aims

To investigate the temperature rise in a known mass of water by the combustion of methanol, ethanol, propanol and butanol.

You can compare different alcohols if you control factors such as the mass or volume of water.

Apparatus

- eye protection
- measuring cylinder
- calorimeter
- stand, boss and clamp
- thermometer
- stirrer
- spirit lamps containing alcohols
- ±0.01 g balance.

A **calorimeter** is a metal can, usually made of copper, used to contain water in these experiments. You could use a clean empty food can instead.

Methanol is toxic, so you should take care to avoid skin contact. The alcohols are flammable.

Method

1. Securely clamp the calorimeter.
2. Add a known volume of cold water to the calorimeter, then record its temperature.
3. Measure and record the mass of a spirit lamp, including its fuel and lid.
4. Place the spirit lamp underneath the calorimeter, at a height where you can comfortably remove and replace its lid.
5. Remove the lamp's lid, and light the wick.
6. Stir the water as it is heated. When the temperature has increased sufficiently, replace the lid to put out the flame.
7. Measure and record the water temperature, and the mass of the spirit lamp with lid.
8. Empty the calorimeter and repeat steps 2 to 7 with the other alcohols.

thermometer
stirrer
clamp
calorimeter
water
spirit lamp
ethanol

You could aim for a given temperature rise, e.g. 20 °C, or heat for a given time.

Results

Alcohol	Temperature (°C)		Mass of lamp (g)	
	start	end	start	end

You should record your results for each alcohol on a suitable table. Remember to record the mass or volume of water used.

Analysis

For each alcohol, calculate:

1. the increase in temperature
2. the decrease in mass (due to alcohol burning)
3. answer 1 ÷ answer 2 (°C/g of alcohol burned).

The 'better' the fuel is, the greater the temperature rise per gram of fuel used.

Now try this

A student uses spirit lamps and a calorimeter to investigate the use of four alcohols as fuels. State **four** factors, other than the alcohol and the mass of water used, that could affect the temperature rise observed. **(4 marks)**

Nanoparticles

Nanoparticles are structures consisting of only a few hundred atoms.

Small size

Nanoparticles have a very small size, so

- **nanoparticulate** materials have different properties from the same substance in larger pieces as a **bulk material**.

This makes nanoparticles useful for:

- 👍 sunscreens – they still absorb harmful ultraviolet light from the Sun but cannot be seen on the skin

- 👍 lightweight strong materials, such as carbon nanotubes in tennis rackets

- 👍 future drug delivery systems – buckyballs consist of hollow balls of carbon atoms.

Surface area to volume ratio

Nanoparticles have a very small size, so:

- they have a very large **surface area:volume ratio**.

This makes them useful as:

- 👍 catalysts, e.g. as coatings for self-cleaning surfaces and clothes.

Hazards and risks

There are possible hazards associated with the use of nanoparticles. They may:

- 👎 be breathed in, absorbed through the skin or transported into cells

- 👎 take a long time to break down

- 👎 attract toxic substances to their surfaces.

The risks to health and the environment may be difficult to predict and to measure.

📷 Maths skills **Calculating surface area:volume ratio**

You can calculate the surface area, volume, and surface area:volume ratio of regularly shaped objects. For example, imagine a cube-shaped nanoparticle with sides of 1 nm:

1 nm
1 nm
1 nm

✓ surface area of one face = (1 nm) × (1 nm)
$$= 1 \text{ nm}^2$$

total surface area = 6 × 1 nm² = 6 nm²

✓ volume of cube = (1 nm) × (1 nm) × (1 nm)
$$= 1 \text{ nm}^3$$

Ignoring the units for area and volume:

$$\text{surface area to volume ratio} = \frac{\text{total surface area}}{\text{volume}} = \frac{6}{1}$$

This gives a ratio of 6:1.

Worked example

Which of the objects is about the same size as a typical nanoparticle? **(1 mark)**

	Object	Size (nm)
☐ A	gold atom	0.4
☒ B	egg white molecule	10
☐ C	bacterium	1000
☐ D	yeast cell	4000

You need to be able to compare the size of nanoparticles with the sizes of atoms and molecules.

Nanoparticles are between 1 nm and 100 nm in size. Remember that 1 nm is one nanometre – one-millionth of a millimetre or 1×10^{-9} m.

Now try this

1 Scientists have developed glass coated with particulate vanadium dioxide. This coating is invisible to the naked eye but it significantly reduces the transfer of heat energy through the glass. Suggest two advantages of using this type of glass for windows. **(2 marks)**

Bulk materials

The properties of a material make it suitable for some uses but not for others.

Some general properties

glass ceramics
- transparent, hard but brittle
- poor conductors of heat and electricity

window glass, bottles

polymers
- transparent → translucent → opaque
- poor conductors of heat and electricity
- often tough and ductile

bottles, crates, carrier bags

Materials

clay ceramics
- opaque, hard but brittle
- poor conductors of heat and electricity

bricks, china and porcelain

metals
- can be polished to a shine
- good conductors of heat and electricity
- hard, tough and ductile

cars, bridges, electrical cables

An example

Glass is used for windows because it is:

👍 transparent, hard and a poor conductor of heat.

Unfortunately, glass is also:

👎 brittle.

Perspex° is a polymer that could be used for windows instead because it is:

👍 transparent and tough.

However, Perspex° also:

👎 scratches more easily.

Some key words

transparent	clear and fully see-through
translucent	lets light through but not detailed shapes
opaque	does not let light through
brittle	will crack or break when hit
tough	resists cracking and breaking
ductile	can be bent, twisted or stretched without cracking or breaking

Worked example

A concrete beam cracks when sufficient force is applied to it. Explain why a much greater force is needed to crack a concrete beam reinforced with steel rods.   **(3 marks)**

Concrete is strong in compression but weak in tension, whereas steel is strong in tension. By combining the two different materials, a composite material is produced that is strong in both compression and tension. This means that it resists bending and cracking better than concrete alone.

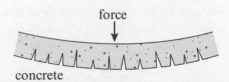

force

concrete

force

steel-reinforced concrete

A composite material has two or more materials combined together, each with different properties. Its overall properties are improved compared with the individual materials it contains.

Now try this

1 Electrical wires consist of a copper core surrounded by a coating of PVC, a polymer. Explain this choice of materials.   **(4 marks)**

Extended response – Materials

There will be one or more 6-mark question on your exam paper. For these questions, you will need to think scientifically and structure your answer logically, showing how the points you make are related to each other.

You can revise the topics for this question, which is about **nanoparticles** and **materials**, on pages 57, 101, 110 and 111.

Worked example

Silver can kill bacteria. Small amounts of silver nanoparticles are used in some socks, plasters, kitchen chopping boards and washing machines. However, some people are allergic to silver. If swallowed in large amounts, silver nitrate causes vomiting, diarrhoea and even death.

Bulk silver is safe on the skin, but nanoparticulate silver might pass into the body.

Explain why nanoparticulate silver may have different properties from bulk silver, and evaluate the use of silver nanoparticles in consumer products. **(6 marks)**

Nanoparticles are only 1–100 nm in size, so they may be small enough to pass through the skin and into the blood. They have a much higher surface area than the same amount of ordinary silver, giving them different properties.

Silver is expensive, so nanoparticles mean that the metal can be used more cheaply because less is needed. The nanoparticles kill bacteria, which may prevent illness and keep clothes fresh for longer.

However, silver might cause allergic reactions, and the nanoparticles might cause illness or death if they pass into the body.

Overall, I think the use of silver nanoparticles is justified because there is a greater chance of illness due to harmful bacteria, and it is only used in small amounts such as in plasters.

Command word: Explain

When you are asked to **explain** something, it is not enough just to state or describe it. Your answer must contain some reasoning or justification of the points you make.

Command word: Evaluate

When you are asked to **evaluate** something, you need to review information given to you and then bring it together to form a conclusion. You should include evidence such as strengths and weaknesses, and arrive at a judgement.

 You should give reasons why nanoparticulate materials may have different properties from the same substance in bulk. Two reasons are given here, including how nanoparticles might get into the body without being swallowed.

 You should include advantages, and several are given in this part of the answer.

 The use of 'however' is a good way of moving to the disadvantages.

 Remember to give a justified conclusion.

It is important in this sort of question to organise your answer in a structured way, giving clear lines of reasoning. In particular, you should take care to give advantages and disadvantages, and to come to a supported judgement.

Now try this

A manufacturer of mobile phones is deciding which of two materials, an aluminium alloy or ABS (an addition polymer), to use to make the body of the phone.

Material	Density (g/cm³)	Melting point (°C)	Relative strength	Cost (£/cm³)
aluminium alloy	2.70	660	7	0.40
ABS	1.07	105	1	0.16

Think about the properties that a mobile phone body should have. What properties, in addition to those given, would it need and why?

Evaluate the use of these two materials to make a mobile phone casing. **(6 marks)**

Answers

Extended response questions

In your exam, your answers to 6-mark questions will be marked on how well you present and organise your response, not just on the scientific content. Your responses should contain most or all of the points given in the answers below, but you should also make sure that you show how the points link to each other, and structure your response in a clear and logical way.

1. Formulae

1 A molecule consists of two bromine atoms **(1)** chemically joined together. **(1)**

2 $CuCO_3$ **(1)**

   No marks if 3 is shown as superscript or element symbols do not start with capital letters.

3 Two from the following for 1 mark each:
   - It contains two oxygen atoms and two hydrogen atoms for every magnesium atom.
   - It has three different elements/five atoms in total, chemically joined together.
   - It contains one magnesium ion and two hydroxide ions.

2. Equations

1 (a) $2Mg + O_2 \rightarrow 2MgO$ **(1)**

   (b) $N_2 + 3H_2 \rightarrow 2NH_3$ **(1)**

   (c) $CH_4 + 2O_2 \rightarrow CO_2 + 2H_2O$ **(1)**

2 (a) $CuO(s) + 2HNO_3(aq) \rightarrow Cu(NO_3)_2(aq) + H_2O(l)$ (1 mark for balancing, 1 mark for state symbols)

   (b) $2Fe(s) + 3Cl_2(g) \rightarrow 2FeCl_3(s)$ (1 mark for balancing, 1 mark for state symbols)

3. Ionic equations

1 $Pb^{2+} + 2Br^- \rightarrow PbBr_2$ **(1)**

2 (a) $BaCl_2(aq) + Na_2SO_4(aq) \rightarrow BaSO_4(s) + 2NaCl(aq)$ **(1)**

   (b) $Ba^{2+}(aq) + SO_4^{2-}(aq) \rightarrow BaSO_4(s)$ **(1)**

4. Hazards, risks and precautions

1 one from: to warn about the dangers; to let people know about precautions to take; for people who cannot read **(1)**

2 chlorine: use a fume cupboard/open the windows **(1)**
   hydrogen: avoid naked flames **(1)**

5. Atomic structure

1 central nucleus **(1)** containing protons and (usually) neutrons **(1)**
   surrounded by electrons **(1)** in shells **(1)**

2 Atoms contain equal numbers of protons and electrons. **(1)**
   Protons have a +1 charge and electrons have –1 charge (so charges cancel out). **(1)**

3 $(9.1094 \times 10^{-31})/(1.6726 \times 10^{-27})$ **(1)**
   $= 5.446 \times 10^{-4}$ or $\frac{1}{1836}$ **(1)**

6. Isotopes

1 Each atom has the same number of protons and electrons **(1)**, 35 protons and electrons **(1)**, different numbers of neutrons **(1)**, 44 or 46 neutrons **(1)**.

2 $(35 \times 75.8) + (37 \times 24.2) = 2653 + 895.4 = 3548.4$ **(1)**
   $A_r = \frac{3548.4}{100} = 35.484 = 35.5$ to 1 decimal place **(1)**

7. Mendeleev's table

1 in order of increasing atomic mass **(1)**, taking into account the properties of the elements and their compounds **(1)**

2 one from: it had gaps/there were pair reversals/some groups contained metals and non-metals **(1)**

8. The periodic table

1 Both relative atomic masses are 59 when rounded to the nearest whole number. **(1)**

9. Electronic configurations

1 diagram with three circles **(1)** with two electrons in the first circle, eight in the second and seven in the outermost circle **(1)**, e.g.

   You can show electrons as dots or as crosses.

2 number of occupied shells is the same as the period number **(1)**; the number of electrons in the outer shell is the same as the group number (except for the elements in group 0, which have full outer shells). **(1)**

10. Ions

1 (a) Li^+ **(1)**

   (b) Mg^{2+} **(1)**

   (c) S^{2-} **(1)**

   (d) Br^- **(1)**

2 a charged particle **(1)** formed when an atom or group of atoms loses or gains electrons. **(1)**

3 (a) protons = 20 **(1)**
      neutrons = $(40 - 20) = 20$ **(1)**
      electrons = $(20 - 2) = 18$ **(1)**

   (b) protons = 9 **(1)**
      neutrons = $(19 - 9) = 10$ **(1)**
      electrons = $(9 + 1) = 10$ **(1)**

11. Formulae of ionic compounds

1 (a) CaS **(1)**

   (b) $FeCl_2$ **(1)**

   (c) NH_4OH **(1)**

   (d) $(NH_4)_2CO_3$ **(1)**

   (e) Na_2SO_4 **(1)**

12. Properties of ionic compounds

1 (a) Its ions are not free to move/are held in fixed positions (in the lattice). **(1)**

   (b) Heat it **(1)** until it melts/becomes molten. **(1)**

   Remember: aluminium oxide is insoluble, so you cannot dissolve it in water.

   (c) Aluminium oxide has a high melting point **(1)** so a lot of energy is needed. **(1)**

13. Covalent bonds

1 (strong) bond formed when a pair of electrons is shared **(1)** between two atoms **(1)**

2 dot-and-cross diagram for oxygen, e.g.

two overlapping circles with two dots and two crosses in the overlap **(1)**

two pairs of non-bonding electrons in each oxygen atom **(1)**

14. Simple molecular substances

1 There are weak intermolecular forces between ammonia molecules **(1)**, which are easily overcome/need little energy to break. **(1)**

2 The intermolecular forces between petrol molecules and water molecules are weaker **(1)** than those between petrol molecules **(1)** and between water molecules. **(1)**

15. Giant molecular substances

1 the following points for 1 mark each to a maximum of 4 marks:

similarities:

- only contain carbon atoms
- atoms covalently bonded to each other
- lattice structure.

differences:

- Diamond has four covalent bonds per atom *and* graphite has three.
- Diamond has no delocalised electrons *and* graphite does.
- Graphite has a layered structure (with weak intermolecular forces) *and* diamond does not.

2 Graphite has weak intermolecular forces **between its layers (1)** so the layers can slide over each other. **(1)**

16. Other large molecules

1 two from the following for 1 mark each: almost transparent/flexible/conducts electricity

2 Buckminsterfullerene is not a giant molecular substance **(1)**, weak intermolecular forces are overcome on heating **(1)**, but there are many strong covalent bonds to break in diamond. **(1)**

17. Metals

1 (a) Layers of positive ions/atoms **(1)** can slide over each other. **(1)**

(b) Delocalised electrons/free electrons/sea of electrons **(1)** can move through the structure. **(1)**

2 (It should conduct electricity) because it will have delocalised electrons. **(1)**

18. Limitations of models

1 One limitation of each model:

- Empirical formula does not show: actual number of atoms/how the atoms are arranged/three-dimensional shape of molecule/bonding and non-bonding electrons/sizes of atoms relative to their bonds. **(1)**
- Molecular formula does not show: how the atoms are bonded/three-dimensional shape of molecule/bonding and non-bonding electrons/sizes of atoms relative to their bonds. **(1)**
- Structural formula does not show: three-dimensional shape of molecule/bonding and non-bonding electrons/sizes of atoms relative to their bonds. **(1)**
- Drawn formula does not show: three-dimensional shape of molecule/bonding and non-bonding electrons/sizes of atoms relative to their bonds. **(1)**

- Ball-and-stick model does not show: element symbols/bonding and non-bonding electrons/sizes of atoms relative to their bonds. **(1)**
- Space-filling model does not show: element symbols/bonding and non-bonding electrons/some atoms in complex models. **(1)**
- Dot-and-cross diagram does not show: three-dimensional shape of molecule/sizes of atoms relative to their bonds. **(1)**

19. Relative formula mass

1 (a) 18 **(1)**

(b) 44 **(1)**

(c) 40 **(1)**

(d) 154 **(1)**

(e) 134.5 **(1)**

(f) 142 **(1)**

(g) 78 **(1)**

(h) 234 **(1)**

20. Empirical formulae

1 (a) Mg: $\frac{0.36}{24} = 0.015$ and O: $\frac{0.24}{16} = 0.015$ **(1)**

Divide by 0.015

Mg: $\frac{0.015}{0.015} = 1$ and O: $\frac{0.015}{0.015} = 1$ **(1)**

Empirical formula is MgO. **(1)**

(b) Air/oxygen must be let in (to react with the magnesium). **(1)**

Magnesium/magnesium oxide must not be allowed to escape. **(1)**

21. Conservation of mass

1 M_r of $CH_4 = 12 + (4 \times 1) = 16$ **(1)**

M_r of $O_2 = (2 \times 16) = 32$ **(1)**

10 g of CH_4 reacts with $\left(2 \times \frac{32}{16}\right) \times 10$ g of O_2 **(1)**

$= 40$ g **(1)**

2 $2K + Cl_2 \rightarrow 2KCl$ **(1)**

A_r of K = 39 and M_r of KCl = 39 + 35.5 = 74.5 **(1)**

20 g of KCl is produced from $\left(\frac{39}{74.5}\right) \times 20$ g of K **(1)**

$= 10.5$ g or 10.47 g **(1)**

22. Reacting mass calculations

1 The reactant not in excess is the limiting reactant **(1)**. When all its particles have been used up, the reaction stops/no product forms. **(1)**

2 amount of $CO_2 = \frac{3.3}{44} = 0.075$ (mol) **(1)**

amount of $H_2 = \frac{0.30}{2} = 0.15$ (mol) **(1)**

ratio of $CO_2:H_2 = 0.075:0.15 = 1:2$ **(1)**

right-hand side is: $\rightarrow CO_2 + 2H_2$

equation is: $C + 2H_2O \rightarrow CO_2 + 2H_2$ **(1)**

23. Concentration of solution

1 (a) $\frac{0.40}{0.50} = 0.80$ g dm^{-3} **(1)**

(b) 100 cm$^3 = \frac{100}{1000} = 0.100$ dm^3 **(1)**

$\frac{1.25}{0.100} = 12.5$ g dm^{-3} **(1)**

2 150 cm$^3 = \frac{150}{1000} = 0.150$ dm^3 **(1)**

mass = concentration × volume

mass = 40 × 0.150 = 6.0 g **(1)**

24. Avogadro's constant and moles

1. (a) $\frac{22.5}{18} = 1.25$ mol (1)

   (b) $3 \times 1.25 = 3.75$ mol (1)

   (c) $3.75 \times 6 \times 10^{23} = 2.25 \times 10^{24}$ (1)

2. (a) amount $= \frac{6.0}{12} = 0.50$ mol (1)

   number $= 0.50 \times 6.02 \times 10^{23} = 3.0 \times 10^{23}$ atoms (1)

   (b) amount $= \frac{1.00 \times 10^{12}}{6.02 \times 10^{23}} = 1.66 \times 10^{-12}$ mol (1)

   mass $= 12 \times 1.66 \times 10^{-12} = 2.0 \times 10^{-11}$ g (1)

25. Extended response – Types of substance

Answer could include the following points (6):

- Mercury contains positively charged mercury ions.
- It is not in a regular lattice.
- There is a sea of delocalised electrons.
- The electrons are free to move and carry charge.
- Liquid zinc chloride contains oppositely charged particles.
- They are attracted to each other but are not in a regular lattice.
- The ions are free to move and carry charge.
- Paraffin oil molecules are attracted to each other by weak intermolecular forces.
- The molecules are free to move.
- The molecules are uncharged so paraffin oil does not conduct electricity.

26. States of matter

1. change of state: sublimation (1)

   Forces of attraction (between particles) are overcome (as energy is transferred to the particles). (1)

   Arrangement changes from regular to random. (1)

   Arrangement changes from close together to far apart. (1)

   Particles no longer just vibrate about fixed positions but can move quickly in all directions. (1)

27. Pure substances and mixtures

1. A pure substance contains only one element or compound (1), but mineral water will be a mixture (1) of water and other substances. (1)

2. 18-carat gold must be a mixture (1) because pure substances have a sharp melting point/mixtures melt over a range of temperatures. (1)

28. Distillation

1. (a) to cool the vapour (1); to turn it from the gas state to the liquid state (1)

   (b) (different) boiling points (1)

   (c) hottest at the bottom and coldest at the top/gets cooler towards the top (1)

29. Filtration and crystallisation

1. Answer should include the following points for 1 mark each to a maximum of 4 marks:

   - Add water to dissolve the copper chloride.
   - Filter to remove the glass and dust (as a residue).
   - Heat the filtrate to produce copper chloride crystals (by crystallisation).
   - Pat the copper chloride crystals dry with filter paper/put them in a warm oven.

30. Paper chromatography

1. A 0.2 (1) E 0.9 (1)

2. The substance forms stronger attractive forces with the solvent/mobile phase (1) than with the paper/stationary phase. (1)

31. Core practical – Investigating inks

Answer could include the following points for 1 mark each to a maximum of 2 marks:

- Clamp the boiling tube.
- Use a blue flame with the air hole half open.
- Do not heat to dryness.
- Put the test tube in a beaker of cold water/iced water.

32. Drinking water

1. safe to drink (1)

2. Sedimentation removes **large** insoluble particles (1); filtration removes **small** insoluble particles (1); chlorination kills microbes/sterilises the water (1).

33. Extended response – Separating mixtures

Answer could include the following points (6):

- Add water to the mixture.
- Stir the mixture.
- Sodium carbonate should dissolve but calcium carbonate should not.
- This is because their solubility is different/calcium carbonate is (almost) insoluble.
- Filter the mixture.
- Calcium carbonate stays behind as a residue.
- Wash the residue in the filter paper (to remove any sodium carbonate).
- Leave washed residue in a warm place to dry.
- Filtrate is sodium carbonate solution.
- Heat to evaporate (most of) the water.
- Leave solution to cool/crystallise sodium carbonate.
- Filter/decant to separate crystals.
- Leave crystals in a warm place to dry.

34. Acids and alkalis

1. Sodium chloride solution is neutral but hydrogen chloride solution is acidic. (1)

   Sodium chloride does not release hydrogen ions when it dissolves but hydrogen chloride does. (1)

   Hydrogen ions make the solution acidic. (1)

2. The indicator turns pink (1) because the solution is alkaline, containing hydroxide ions. (1)

3. 1 mark for each correct row to 2 marks maximum:

	Acidic solution	Alkaline solution
Red litmus	(stays) red	(turns) blue
Blue litmus	(turns) red	(stays) blue

35. Strong and weak acids

1. The concentration of hydrogen ions (1) is the same. (1)

2. 0.002 mol dm^{-3} (1)

36. Bases and alkalis

1. An alkali is a soluble base (1) but most bases are insoluble (so are not alkalis). (1)

2. NaCl – sodium chloride (1); $Cu(NO_3)_2$ – copper nitrate (1); $MgCl_2$ – magnesium chloride (1); $Al_2(SO_4)_3$ – aluminium sulfate (1); $CaCl_2$ – calcium chloride (1)

37. Core practical – Neutralisation

(a) two precautions for precise measurements for 1 mark each to give a total of 2 marks, e.g.
  - Use a measuring cylinder with a capacity similar to the volume being used/use a (volumetric) pipette.
  - Use a balance of 2 decimal places (±0.01 g).
  - Use a pH meter.
  - Calibrate the pH meter.

(b) Add a spot of liquid to the universal indicator paper. **(1)**
Match the colour to the pH colour chart. **(1)**
Work carefully, e.g. dip a glass rod into the reaction mixture/put the indicator paper on a white tile/wait 30 s for colour to develop. **(1)**

(c) The pH will be 0–6 at the start **(1)** because the solution contains excess $H^+(aq)$ ions. **(1)**
The pH will be 7 at some point **(1)** because the acid will be neutralised by the base/alkali. **(1)**
The pH will be 8–14 at the end **(1)** because the solution contains excess $OH^-(aq)$ ions. **(1)**

38. Core practical – Salts from insoluble bases

(a) during the reaction: bubbling **(1)**; the white solid disappearing **(1)**
when calcium carbonate in excess: a white solid mixed with the liquid/cloudy mixture **(1)**

(b) 1 mark for each of the following points to 6 marks maximum:
  - put some dilute nitric acid into a beaker/suitable container
  - add calcium carbonate
  - add until bubbling stops/excess solid is seen
  - filter (to remove excess solid)
  - heat the filtrate
  - this will remove some of the water
  - leave to cool
  - decant/pour away/filter to remove the excess liquid
  - leave crystals in a warm place/dry with paper

39. Salts from soluble bases

1 Filter the mixture **(1)** so the insoluble powder and indicator stay behind (and the salt solution passes through). **(1)**

40. Making insoluble salts

1 (a) calcium carbonate **(1)**
  (b) no precipitate **(1)**
  (c) lead sulfate **(1)**

41. Exam skills – Making salts

Answer could include the following points **(6)**:
- Suitable solutions: lead(II) nitrate and sodium chloride/potassium chloride/ammonium chloride.
- Lead compounds are toxic.
- Wear eye protection.
- Wear gloves.
- $Pb(NO_3)_2(aq) + 2NaCl(aq) \rightarrow PbCl_2(s) + 2NaNO_3(aq)$
- $Pb(NO_3)_2(aq) + 2KCl(aq) \rightarrow PbCl_2(s) + 2KNO_3(aq)$
- $Pb(NO_3)_2(aq) + 2NH_4Cl(aq) \rightarrow PbCl_2(s) + 2NH_4NO_3(aq)$
- Mix the two solutions.
- Filter the mixture.
- Lead(II) chloride stays behind as a residue.
- Wash the residue in the filter paper.
- Use water/distilled water.
- This works because lead(II) chloride is insoluble.
- Leave washed residue in a warm place to dry.

42. Electrolysis

1 (a) potassium (at the negative electrode) **(1)**; iodine (at the positive electrode) **(1)**
  Note that iodide is incorrect.

  (b) negative electrode: $K^+ + e^- \rightarrow K$ **(1)**
  positive electrode: $2I^- \rightarrow I_2 + 2e^-$ **(1)**

  (c) Oxidation occurs at the positive electrode **(1)** because iodide ions lose electrons. **(1)**

43. Electrolysing solutions

1 (a) H^+ from water and acid **(1)**; OH^- from water **(1)**; SO_4^{2-} from acid **(1)**

  (b) hydrogen from the cathode/negative electrode **(1)**; $2H^+(aq) + 2e^- \rightarrow H_2(g)$ **(1)**; oxygen at the anode/positive electrode **(1)**; $4OH^-(aq) \rightarrow 2H_2O(l) + O_2(g) + 4e^-$ **(1)**

44. Core practical – Investigating electrolysis

(a) Ethanol evaporates more quickly/has a lower boiling point than water. **(1)**

(b) 1 mark for each of the following points to 2 marks maximum:
  - Some copper did not stick to the cathode/fell off during the experiment.
  - Some copper fell off the cathode during washing/drying/weighing.
  - Some metal fell off the anode during the experiment.
  - The anode was not pure copper/contained other metals.

45. Extended response – Electrolysis

Answer could include the following points **(6)**:
dissolving in molten potassium chloride:
- Ions are not free to move in solid lithium chloride.
- Ions must be free to move for electrolysis to happen.
- Ions will be free to move in liquid/molten potassium chloride.
cathode process:
- Lithium ions are attracted to the cathode.
- Lithium ions gain electrons and become lithium atoms.
- $Li^+ + e^- \rightarrow Li$
- This is reduction because the ions gain electrons.
anode process:
- Chloride ions are attracted to the anode.
- Chloride ions lose electrons and become chlorine atoms.
- Covalent bond forms between two chlorine atoms.
- There is a shared pair of electrons.
- Pairs of chlorine atoms form chlorine molecules.
- $2Cl^- \rightarrow Cl_2 + 2e^-$
- This is oxidation because the ions lose electrons.

46. The reactivity series

1 (a) magnesium/calcium **(1)**
  (b) iron **(1)** and zinc **(1)**

47. Metal displacement reactions

1 (a) magnesium, zinc, iron, copper **(1)** This is in order of decreasing change in temperature **(1)**

  (b) 0.0 °C/no change in temperature **(1)** This is because silver cannot displace copper from copper sulfate solution/no reaction **(1)**

48. Explaining metal reactivity

1 Aluminium is more reactive than iron **(1)** because it can displace iron from iron(III) oxide. **(1)** This is because it has a greater tendency to lose electrons/form cations. **(1)**

49. Metal ores

1 Copper is more resistant to corrosion **(1)** because it is lower in the reactivity series. **(1)**

2 Oxidation is a gain of oxygen. **(1)** Reduction is a loss of oxygen. **(1)**

3 Platinum is low in the reactivity series/below gold/very unreactive **(1)**, so it does not combine with other elements. **(1)**

50. Iron and aluminium

1 Tin is less reactive than carbon/carbon is more reactive than tin. **(1)**

Calcium is more reactive than carbon/carbon is less reactive than calcium. **(1)**

So carbon can reduce tin oxide to tin, but not calcium oxide to calcium. **(1)**

2 Electrolysis uses electricity. **(1)** This is expensive/more expensive than using carbon. **(1)**

51. Biological metal extraction

1 Plants absorb metal compounds through their roots. **(1)** The metal compounds become concentrated in the plants. **(1)** The plants are harvested and burned. **(1)** The ash contains (high concentrations of) metal compounds. **(1)**

2 advantage: low-grade ores can be used/mining not needed/ happens naturally **(1)**

disadvantage: slow/further processes needed **(1)**

52. Recycling metals

1 two from the following for 1 mark each: less waste rock produced/less carbon dioxide (or named pollutant) produced/ less waste sent to landfill

2 Less energy is needed to melt aluminium than to melt steel. **(1)**

More energy is needed to extract aluminium by electrolysis than is needed to extract iron in the blast furnace. **(1)**

3 Recycling uses metals already extracted from ores **(1)**, so no ore/less ore must be mined to produce new metal. **(1)**

53. Life-cycle assessments

1 The four main stages in a life-cycle assessment are: obtaining raw materials, manufacture, use, disposal. **(1)**

2 Another reason for carrying out a life-cycle assessment is to identify a stage that could be improved on. **(1)**

54. Extended response – Reactivity of metals

Answer could include the following points **(6)**:

expected order of reactivity:

- magnesium > zinc > copper

basic method:

- Add one of the solutions to three test tubes/other suitable container, e.g. spotting tile.
- Add a metal powder to each test tube.
- Observe a change in the solution and/or the metal powder, e.g. a colour change or temperature increase.
- Repeat with each of the other two solutions in turn.

expected results:

- An orange–brown coating of metal is obtained when magnesium or zinc is added to copper sulfate.
- The temperature increases when magnesium or zinc is added to copper sulfate.
- The blue colour of the solution fades when magnesium or zinc is added to copper sulfate.
- There is a black coating of metal when magnesium is added to zinc sulfate solution.
- There is no observed change with other combinations of metal and solution.

explanation:

- Magnesium and zinc are more reactive than copper/can displace copper from its compounds.
- Magnesium and zinc have a greater tendency than copper to lose electrons/are more readily oxidised/better reducing agents.
- $Mg + CuSO_4 \rightarrow MgSO_4 + Cu$
- $Zn + CuSO_4 \rightarrow ZnSO_4 + Cu$
- Magnesium is more reactive than zinc/can displace zinc from its compounds.
- Magnesium has a greater tendency than zinc to lose electrons/ is more readily oxidised/is a better reducing agent.
- $Mg + ZnSO_4 \rightarrow MgSO_4 + Zn$
- Copper is less reactive than magnesium and zinc/cannot displace magnesium and zinc from their compounds.
- Zinc is less reactive than magnesium/cannot displace magnesium from its compounds.

55. Transition metals

1 high melting point **(1)**; high density **(1)**; form coloured compounds **(1)**; catalytic activity **(1)**

2 Copper is a transition metal and forms coloured compounds. **(1)** Aluminium is in group 3/not a transition metal, so it forms white or colourless compounds **(1)**. Therefore, copper(I) oxide must be the red powder and aluminium oxide must be the white powder **(1)**.

56. Rusting

1 Iron gains oxygen **(1)** to form iron oxide **(1)**.
Iron atoms also lose electrons to form iron(III) ions.

2 The paint stops air and water reaching the metal beneath. **(1)** Rusting needs air/oxygen and water **(1)**.

57. Alloys

1 Copper is a similar colour to gold/orange–brown colour. **(1)** It is unreactive/does not corrode easily/is low down on the reactivity series (similar to gold) **(1)**.

58. Extended response – Alloys and corrosion

Answer could include the following points **(6)**:

information from the table:

- The alloy has a (2.8 times) lower density than steel.
- Aircraft parts made from the alloy will be lighter than the same parts made from steel.
- Lighter parts will reduce fuel consumption/other named advantage, e.g. ability to fly.
- The alloy is (4.7 times) stronger than aluminium alone.
- The alloy is (2.5 times) weaker than steel.
- The alloy has a better strength:mass ratio than steel (1.7 compared with 1.5).

information from prior knowledge and understanding:

- Aluminium is a metal.
- Layers of aluminium atoms can slide over each other.
- Metals can be stretched/bent/shaped.
- Different-sized atoms in alloys make it more difficult for layers to slide over each other.

- Aluminium alloy should be harder than aluminium alone.
- Steel corrodes/rusts easily (but aluminium does not).

59. Core practical – Accurate titrations

(a) one of the following points for 1 mark:

The end-point is not known accurately/acid is not added drop by drop near the end-point/too much acid may be added.

(b) The concordant titres are the titres from runs 2 and 3. **(1)**

$$\text{mean} = \frac{22.60 \text{ cm}^3 + 22.70 \text{ cm}^3}{2} = 22.65 \text{ cm}^3 \textbf{ (1)}$$

First mark can be obtained using correct working out.

(c) pink to colourless **(1)**

60. Concentration calculations

1 (a) $\text{volume} = \frac{400}{1000} = 0.400 \text{ dm}^3$ **(1)**

$\text{concentration} = \frac{0.0100}{0.400} = 0.0250 \text{ mol dm}^{-3}$ **(1)**

(b) $M_r = 40 + [2 \times (16 + 1)] = 74$ **(1)**

$\text{concentration} = 0.025 \times 74 = 1.85 \text{ g dm}^{-3}$ **(1)**

2 $\text{volume} = \frac{250}{1000} = 0.250 \text{ dm}^3$ **(1)**

$\text{amount} = 0.100 \times 0.250 = 0.025 \text{ mol}$ **(1)**

$\text{mass} = 0.025 \times 40 = 1.0 \text{ g}$ **(1)**

61. Titration calculations

1 $\text{volume of KOH} = \frac{25.00}{1000} = 0.0250 \text{ dm}^3$

$\text{volume of HNO}_3 = \frac{28.00}{1000} = 0.0280 \text{ dm}^3$

$\text{amount of HNO}_3 = 0.200 \times 0.0280 = 0.00560 \text{ mol}$ **(1)** *allow 5.60×10^{-3} mol*

1:1 mole ratio of KOH to HNO_3, so 0.00560 mol KOH **(1)**

$\text{concentration of KOH} = \frac{0.00560}{0.0250} = 0.224 \text{ mol dm}^{-3}$ **(1)**

2 $\text{volume of NaOH} = \frac{25}{1000} = 0.025 \text{ dm}^3$

$\text{amount of NaOH} = 0.20 \times 0.025 = 0.0050 \text{ mol}$ **(1)**

1:1 mole ratio of NaOH to HCl, so 0.0050 mol HCl **(1)**

$\text{volume of HCl} = \frac{0.0050}{0.50} = 0.010 \text{ dm}^3$ **(1)** *allow 10 cm³*

62. Percentage yield

1 $\frac{360}{400} \times 100 = 90\%$ **(1)**

2 incomplete reactions **(1)**; practical losses during the experiment **(1)**; side reactions/competing unwanted reactions. **(1)**

63. Atom economy

1 (a) M_r of $CO_2 = 44$ and M_r of $H_2 = 2$ **(1)**

total M_r of desired products $= 2 \times 2 = 4$

total M_r of all products $= 44 + 4 = 48$

$\text{atom economy} = \frac{4}{48} \times 100$ **(1)**

$= 8.33\%$ **(1)**

(b) (i) M_r of $H_2 = 2$ and M_r of $O_2 = 32$ **(1)**

total M_r of desired products $= 2 \times 2 = 4$

total M_r of all products $= 4 + 32 = 36$

$\text{atom economy} = \frac{4}{36} \times 100$ **(1)**

$= 11.1\%$ **(1)**

(ii) The atom economy could be improved by selling the oxygen (e.g. for medical uses). **(1)**

64. Molar gas volume

1 $24\,000 \times 0.25 = 6000 \text{ cm}^3$ **(1)**

2 $\frac{1.2}{24} = 0.05 \text{ mol}$ **(1)**

3 $40 \times \frac{3}{2}$ **(1)** $= 60 \text{ dm}^3$ **(1)**

3 mol of H_2 needed for 2 mol of N_2, so 3/2 mol of H_2 needed for each mol of N_2

65. Gas calculations

1 $\text{amount of } H_2 = \frac{120}{24\,000} = 0.0050 \text{ mol } (5 \times 10^{-3} \text{ mol})$ **(1)**

$\text{amount of Na} = 2 \times 0.0050 = 0.010 \text{ mol}$ **(1)**

$\text{mass of Na} = 23 \times 0.010 = 0.23 \text{ g}$ **(1)**

66. Exam skills – Chemical calculations

1 (a) $\text{volume} = \frac{100}{1000} = 0.1 \text{ dm}^3$ **(1)**

$\text{concentration} = \frac{11.7}{0.1} = 117 \text{ g dm}^{-3}$ **(1)**

(b) M_r of NaCl $= 23 + 35.5 = 58.5$ **(1)**

$\text{concentration} = \frac{117}{58.5} = 2 \text{ mol dm}^{-3}$ **(1)**

(c) $\frac{11.7}{58.5} = 0.2 \text{ mol}$ **or** $2 \times 0.1 = 0.2 \text{ mol}$ **(1)**

(d) $\frac{0.2}{2}$ **(1)** $= 0.1 \text{ mol}$ **(1)**

(e) M_r of $Cl_2 = 2 \times 35.5 = 71$ **(1)**

$\text{mass} = 0.1 \times 71 = 7.1 \text{ g}$ **(1)**

(f) total M_r of all products $= (2 \times 40) + 2 + 71 = 153$ **(1)**

$\text{atom economy} = \frac{71}{153} \times 100$ **(1)** $= 46.4\%$ **(1)**

67. The Haber process

1 (a) The symbol means that the reaction is reversible. **(1)**

(b) Add the substance to be tested to anhydrous copper sulfate. **(1)** If the copper sulfate turns blue, water is present. **(1)**

68. More about equilibria

1 (a) The equilibrium concentration is decreased **(1)** because the position of equilibrium moves to the left/in the direction of the greatest number of molecules of gas. **(1)**

(b) The equilibrium concentration is increased **(1)** because the position of equilibrium moves to the right/away from the endothermic reaction/in the direction of the exothermic reaction. **(1)**

(c) The equilibrium concentration is increased **(1)** because the position of equilibrium moves to the right/away from the substance increased in concentration. **(1)**

69. Making fertilisers

1 (a) (i) ammonia solution **(1)** and dilute sulfuric acid **(1)**

(ii) natural gas **(1)**, air **(1)**, water **(1)** and sulfur **(1)**

(b) two comparisons from the following for 1 mark each:

Laboratory scale	Industrial scale
few stages	several stages
small scale	large scale
batch process	continuous process

70. Fuel cells

1 advantage: environment, peaceful roads not so noisy/car quiet inside **(1)**

disadvantage: difficult to hear car coming/may be more pedestrian accidents **(1)**

71. Extended response – Reversible reactions

Answer could include the following points (6):

catalyst:

- A catalyst should be used.
- A catalyst does not affect the equilibrium yield of hydrogen, but it:
  - reduces the time needed to reach equilibrium
  - increases the rate of the forward and backward reactions
  - increases these rates by the same amount/ratio.

temperature:

- A high temperature should be used.
- Equilibrium will be reached quickly.
- Equilibrium yield of hydrogen will be high.
- As the temperature increases, the position of equilibrium moves to the right and more hydrogen is formed.

pressure:

- A low pressure should be used:
  - to obtain a high equilibrium yield of hydrogen
  - because there are more molecules of gas on the right.
- As the pressure increases, the position of equilibrium moves to the left:
  - so less hydrogen is formed at higher pressures.
- However, the rate of reaction would be lower if the pressure is too low:
  - so a compromise pressure may be needed.

72. The alkali metals

1 (a) $2Cs + 2H_2O \rightarrow 2CsOH + H_2$

(1 mark for correct symbols and formulae; 1 mark for correct balancing)

(b) two from the following for 1 mark each:
- violent/very violent reaction explosion
- sparks/flames
- hydrogen/bubbles produced
- metal disappearing

(c) two from the following for 1 mark each:
- good conductors of heat/electricity
- shiny when freshly cut
- soft
- relatively low melting points

73. The halogens

1 (a) The elements get darker going down the group. **(1)**

(b) melting point in the range 250 °C to 350 °C **(1)**, justification, e.g. extrapolated melting points from chlorine to iodine **(1)**

Accepted estimate for melting point is 302 °C.

74. Reactions of halogens

1 The outer shell in iodine is further from the nucleus/more shielded by inner electrons than it is in fluorine. **(1)**

Iodine gains electrons less easily than fluorine does. **(1)**

2 $2Al + 3I_2 \rightarrow 2AlI_3$

(1 mark for correct symbols and formulae; 1 mark for correct balancing)

75. Halogen displacement reactions

1 (a) balanced equation: $Br_2(aq) + 2KI(aq) \rightarrow 2KBr(aq) + I_2(g)$

(1 mark for correct symbols and formulae; 1 mark for correct balancing)

ionic equation: $Br_2(aq) + 2I^-(aq) \rightarrow 2Br^-(aq) + I_2(aq)$

(1 mark for correct symbols and formulae; 1 mark for correct balancing)

(b) Bromine atoms (in Br_2) gain electrons and are reduced to bromide ions. **(1)**

$Br_2 + 2e^- \rightarrow 2Br^-$ **(1)**

Iodide ions lose electrons and are oxidised to iodine. **(1)**

$2I^- \rightarrow I_2 + 2e^-$ **(1)**

(c) Astatine will not react with potassium iodide solution because:
- reactivity decreases down group 7 **(1)**
- astatine is less reactive than iodine. **(1)**

76. The noble gases

1 The densities of helium and neon are less than the density of air **(1)** but the densities of argon and krypton are greater than the density of air **(1)**.

2 It is inert/non-flammable. **(1)** It is denser than air so it sinks and excludes air. **(1)**

3 The electronic configuration is 2. **(1)** Its outer shell is full **(1)**, so it has no tendency to lose/gain/share electrons **(1)**.

77. Extended response – Groups

Answer could include the following points (6):

experiment:

- Put potassium chloride solution in three test tubes.
- Add a few drops of halogen solution to each one.
- Observe any changes.
- Record in a suitable table.
- Repeat with potassium bromide and potassium iodide.

precautions:

- Wear eye protection because solutions are irritants.
- Avoid contact with skin because solutions are irritants.
- Avoid breathing in vapours/keep lab well ventilated because vapours are toxic/harmful.

expected results:

- No changes with potassium chloride.
- Potassium bromide turns darker when chlorine solution is added.
- $Cl_2 + 2KBr \rightarrow 2KCl + Br_2$/bromine is produced.
- Potassium iodide turns darker when chlorine solution is added.
- $Cl_2 + 2KI \rightarrow 2KCl + I_2$/iodine is produced.
- Potassium iodide turns darker when bromine solution is added.
- $Br_2 + 2KI \rightarrow 2KBr + I_2$/iodine is produced.

using the results:

- Chlorine can displace bromine and iodine/oxidise bromide ions and iodide ions.
- Bromine can displace iodine/oxidise iodide ions.
- Iodide cannot displace chlorine or bromine/cannot oxidise chloride or bromide ions.
- Order of reactivity, most reactive to least reactive: chlorine, bromine, iodine.

78. Rates of reaction

1 (a) The rate of reaction increases **(1)**, the surface area to volume ratio of the marble increases **(1)** and so the frequency of collisions increases. **(1)**

(b) The rate of reaction decreases **(1)**, the acid particles become less crowded/there are fewer reactant particles in the same volume **(1)** and so the frequency of collisions decreases. **(1)**

(c) The rate of reaction increases **(1)**, the (acid) particles move around faster/have more energy **(1)**, more collisions have the necessary activation energy or higher energy **(1)** and so the frequency of successful collisions increases. **(1)**

79. Core practical – Investigating rates

Three from the following for 1 mark each:

- same volume of sodium thiosulfate solution
- same concentration of sodium thiosulfate solution
- same amount of cloudiness (measured using the disappearing cross)
- same volume of diluted acid.

These factors also affect the rate of reaction/measured rate of reaction. **(1)**

80. Exam skills – Rates of reaction

(a) 0 s to 45 s: $\frac{43}{45}$ **(1)** = 0.96 cm^3/s **(1)**

45 s to 90 s: $\frac{50-43}{45}$ **(1)** = 0.16 cm^3/s **(1)**

(b) Reactant particles are used up during the reaction **(1)**, so the concentration of hydrogen peroxide decreases **(1)**, the frequency of collisions decreases **(1)** and the rate of reaction decreases. **(1)**

(c) Measure and record the mass of manganese dioxide at the start. **(1)**
(afterwards) Filter **(1)**, dry **(1)**, measure and record the mass of manganese again. **(1)**

81. Heat energy changes

1 1 mark for each correct column (0 marks if two ticks in a column) to 4 marks, e.g.

	Breaking bonds	Making bonds	Temperature of reaction mixture	
			Increases	Decreases
Exothermic process		✓	✓	
Endothermic process	✓			✓

82. Reaction profiles

1 Activation energy is the minimum energy **(1)** needed to start a reaction/to break the bonds in the reactant particles. **(1)**

2 (a) correct axes and shape of graph **(1)**; reactants and products identified **(1)**; activation energy identified **(1)**; energy change identified **(1)**, e.g.

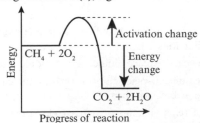

(b) correct axes and shape of graph **(1)**; reactants and products identified **(1)**; activation energy identified **(1)**; energy change identified **(1)**, e.g.

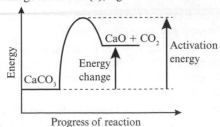

83. Calculating energy changes

1 (a) energy in to break bonds:
    4 × (O–H) = (4 × 464) = 1856 kJ mol^{-1} **(1)**
    energy out when bonds form:
    2 × (H–H) + 1 × (O=O) = (2 × 436) + (1 × 498) **(1)**
    = 872 + 498 = 1370 kJ mol^{-1} **(1)**
    energy change = 1856 – 1370 = +486 kJ mol^{-1} **(1)**

(b) Process is endothermic **(1)** because the energy change is positive/the energy taken in to break bonds is greater than the energy given out when bonds form. **(1)**

2 energy in to break bonds:
    4 × (C–H) + 2 × (O=O) = (4 × 413) + (2 × 498) **(1)**
    = 1652 + 996 = 2648 kJ mol^{-1} **(1)**
    energy out when bonds form:
    4 × (O–H) + 2 × (C=O) = (4 × 464) + (2 × 805) **(1)**
    = 1856 + 1610 = 3466 kJ mol^{-1} **(1)**
    energy change = 2648 – 3466 = –818 kJ mol^{-1} **(1)**

84. Crude oil

1 (a) A hydrocarbon is a compound of carbon and hydrogen **(1)** only. **(1)**

(b) covalent bond **(1)**

2 Crude oil takes millions of years to form/is made extremely slowly **(1)** or is no longer being made. **(1)**

3 chains **(1)**; rings **(1)**

4 (a) (petroleum) gases/petrol/kerosene/diesel oil/fuel oil **(1)**
    not bitumen

(b) polymers/named polymer, e.g. poly(ethene) **(1)**

85. Fractional distillation

1 It is difficult to ignite/not very flammable **(1)**; it has a high viscosity/is very viscous/does not flow easily. **(1)**

2 Oil is evaporated **(1)** and passed into a column, which is hot at the bottom and cool at the top. **(1)** Hydrocarbons (rise), cool and condense at different heights. **(1)**

86. Alkanes

1 three from the following for 1 mark each:

- same general formula
- molecular formulae of neighbouring members differ by CH_2
- gradual variation in physical properties
- similar chemical properties.

2 $C_{21}H_{44}$ **(1)**

3 –40°C (answer in range –35°C to –45°C) **(1)**

87. Incomplete combustion

1 There is a poor supply of air/oxygen **(1)**, so carbon in the fuel is only partially oxidised to carbon monoxide **(1)** or released as carbon particles/soot. **(1)**

2 It is odourless **(1)** and colourless. **(1)**

3 Soot shows that incomplete combustion is happening. **(1)** Carbon monoxide might be forming but soot does not prove this/carbon monoxide is likely to be forming as well (which could be confirmed using a carbon monoxide detector). **(1)**

88. Acid rain

1 two from the following for 1 mark each:

- weathering of buildings/statues
- damage to trees
- harm to living things in river/lakes/soil

2 (a) Hydrocarbon fuels contain impurities of sulfur compounds. **(1)** The sulfur is oxidised to sulfur dioxide when the fuel is used. **(1)**

   (b) Nitrogen and oxygen from the air **(1)** react together in the high temperatures inside an engine. **(1)**

89. Choosing fuels

1 Petrol contains hydrocarbons. **(1)** The carbon in these molecules is oxidised to carbon dioxide. **(1)** Hydrogen does not contain carbon/consists only of hydrogen. **(1)**

2 (a) Crude oil and natural gas are non-renewable. **(1)** They are being used faster than they are formed. **(1)**

   (b) Carbon dioxide may be produced during the manufacture/transport of the fuel. **(1)** For example, fossil fuels are used (in power stations) to generate electricity/to react with steam to make hydrogen. **(1)**

90. Cracking

1 (a) C_2H_4 **(1)**

   (b) This hydrocarbon can be used to make polymers. **(1)**

91. Extended response – Fuels

Answer must include a supported judgement in favour of hydrogen, diesel oil, both or neither. It could include the following points **(6)**:

advantages of hydrogen:
- The only product is water.
- Hydrogen can be produced by the electrolysis of water.
- If water is the raw material, hydrogen could be a renewable resource.
- The electricity needed could be generated using renewable resources (sun, wind, tidal, biomass)
- No carbon dioxide is released when it is used.

disadvantages of hydrogen:
- It is expensive to produce.
- It is difficult to store.
- It may need to be stored under pressure/low temperatures.
- There are few filling stations (so the range is restricted).
- Hydrogen is usually produced from natural gas/coal.
- Carbon dioxide is released in these processes.

advantages of diesel oil:
- It is a liquid (rather than a gas) at room temperature.
- It is easy to store.
- There are many filling stations (so the range is not restricted).

disadvantages of diesel oil:
- Carbon dioxide is also produced.
- Carbon dioxide is a greenhouse gas/linked to global warming/climate change.
- It is produced from crude oil, which is a limited/non-renewable resource.
- It produces carbon particles/soot.
- It produces sulfur dioxide (unless sulfur impurities are removed from fuel before use).
- It produces oxides of nitrogen.
- Sulfur dioxide and oxides of nitrogen cause acid rain.

92. The early atmosphere

1 (a) Carbon dioxide dissolved in the oceans. **(1)**

   (b) Plants produced oxygen **(1)** by photosynthesis. **(1)**

2 Scientists cannot be certain about the Earth's early atmosphere because no measurements were made then/no humans were on Earth then. **(1)**

93. Greenhouse effect

1 (a) carbon dioxide **(1)**, methane **(1)**

   (b) carbon dioxide from burning fossil fuels **(1)**, methane from cattle/livestock/rice paddy field **(1)**

2 Greenhouse gases absorb heat radiated from the Earth. **(1)** The gases then release the heat (into the atmosphere). **(1)**

3 Fossil fuels give off carbon dioxide during combustion. **(1)** Increased consumption releases more carbon dioxide **(1)**, which is a greenhouse gas **(1)**, so the greenhouse effect increases **(1)**.

94. Extended response – Atmospheric science

Answer could include the following points **(6)**:

processes that remove carbon dioxide:
- photosynthesis by plants/algae
- making oxygen and glucose/carbohydrates/starch
- carbon dioxide dissolving in seawater

processes that release carbon dioxide:
- combustion of fossil fuels
- combustion of biomass/plants/trees
- respiration
- volcanic eruptions

observed increase:
- rate of release greater than rate of removal
- increasing use of fossil fuels
- deforestation means less photosynthesis
- limit to the rate at which carbon dioxide can dissolve in the oceans

95. Tests for metal ions

1 Add sodium hydroxide solution. **(1)** Iron(II) chloride forms a green precipitate **(1)**, but iron(III) chloride forms a brown/orange–brown precipitate **(1)**.

2 The different colours interfere with each other. **(1)**

3 The flame test gives a blue–green flame. **(1)** Addition of sodium hydroxide solution gives a blue precipitate. **(1)**

96. More tests for ions

1 Add dilute nitric acid to react with carbonate ions/ions that would interfere with the test **(1)**, then add silver nitrate solution **(1)**. Sodium chloride solution produces a white precipitate of silver chloride; sodium iodide solution produces a yellow precipitate of silver iodide. **(1)**

97. Instrumental methods

1 0.0008 mol dm^{-3} **(1)**

98. Extended response – Tests for ions

Answer could include the following points in a sensible order to determine the identity of the substance **(6)**:

test for carbonate ions:
- Add dilute acid/hydrochloric acid/nitric acid.
- If bubbles of gas form then carbonate ions are present.
- Confirmatory test on gas – limewater turns milky.

test for halide ions:
- Dissolve in water.
- Add dilute nitric acid.
- Add silver nitrate solution.
- Chloride ions give white precipitate.
- Iodide ions give yellow precipitate.

test for metal ions:
- Flame test (on solid or solution).

- Lithium ions give red flame.
- Potassium ions give lilac flame.

99. More about alkanes

1 (a) C_6H_{14} **(1)**

   drawn formula (with all bonds shown) **(1)**, e.g.

   (b) It has only C–C bonds/no C=C bonds. **(1)**

   (c) $C_6H_{14} + 9\frac{1}{2}O_2 \rightarrow 6CO_2 + 7H_2O$ or
        $2C_6H_{14} + 19O_2 \rightarrow 12CO_2 + 14H_2O$

        (1 mark for correct number of products, 1 mark for O_2 correctly balanced)

100. Alkenes

1 (a) Hexene molecules contain carbon and hydrogen **(1)** only **(1)** and the functional group C=C **(1)**.

   (b) Add bromine water. **(1)** There is no change with hexane/the solution stays orange. **(1)** Bromine water is decolourised with hexene/changes from orange to colourless. **(1)**

        Note that 'clear' does not mean 'colourless'.

101. Addition polymers

1 1 mark for the correct polymer structure, 1 mark for use of n on both sides

102. Condensation polymers

1 (a) condensation polymer **(1)**

   (b) The types of monomers needed to manufacture PET are: molecules with two carboxylic acid groups **(1)** and molecules with two alcohol groups. **(1)**

   (c) water **(1)**

103. Biological polymers

1 sugars – starch **(1)**; amino acids – protein **(1)**; nucleotides – DNA **(1)**

104. Polymer problems

1 (a) one advantage from: does not use up landfill sites/useful energy can be obtained **(1)**; one disadvantage from: release of polluting (toxic/harmful/greenhouse) gases/waste of finite resources **(1)**

   (b) one advantage from: does not release polluting (toxic/harmful/greenhouse) gases/does not need complex equipment **(1)**; one disadvantage from: running out of landfill sites/landfill sites take up space/waste does not degrade quickly **(1)**

2 melting and reforming into new objects **(1)**; breaking down into new raw materials **(1)**

3 Poly(ethene) is made from crude oil **(1)**; the cost of crude oil may be higher than the cost of recycling. **(1)**

105. Extended response – Hydrocarbons and polymers

Answer could include the following points in a logical sequence to describe the reactions and to explain how to distinguish between ethane and ethene **(6)**:

ethane:
- C_2H_6
- structure: C–C with six C–H bonds
- saturated
- Complete combustion produces carbon dioxide and water.
- $C_2H_6 + 3\frac{1}{2}O_2 \rightarrow 2CO_2 + 3H_2O$ or
  $2C_2H_6 + 7O_2 \rightarrow 4CO_2 + 6H_2O$

ethene:
- C_2H_4
- structure: C=C with four C–H bonds
- unsaturated
- Complete combustion produces carbon dioxide and water.
- combustion: $C_2H_4 + 3O_2 \rightarrow 2CO_2 + 2H_2O$

differences:
- Less oxygen is needed when ethene burns.
- Less water is produced when ethene burns.
- Ethene reacts with bromine water in an addition reaction/ethane does not.
- $C_2H_4 + Br_2 \rightarrow C_2H_4Br_2$

Note that ethane can *react with bromine in the presence of UV light but in a substitution reaction (not required for GCSE).*

laboratory test:
- Bromine water is decolourised/changes from orange–brown to colourless in ethene.
- no change in colour in ethane

106. Alcohols

1 (a) The members of a homologous series all have the same functional group. **(1)** In alcohols this is the –OH/hydroxyl group (and butanol contains this group). **(1)**

   (b) Drawn formula for butanol, e.g.

   (2 marks if all atoms and bonds are shown correctly, but 1 mark only if OH shown rather than O–H)

   (c) $C_4H_9OH + 6O_2 \rightarrow 4CO_2 + 5H_2O$ or
        $CH_3CH_2CH_2CH_2OH + 6O_2 \rightarrow 4CO_2 + 5H_2O$

        (1 mark for correct formulae; 1 mark for correct balancing)

107. Making ethanol

1 (a) $C_6H_{12}O_6(aq) \rightarrow 2C_2H_5OH(aq) + 2CO_2(g)$

        (1 mark for correct formulae, 1 mark for balancing, 1 mark for state symbols [if formulae are correct])

   (b) enzymes **(1)** from yeast **(1)**

   (c) answer in range 25–35 °C **(1)**

2 Fractional distillation is used **(1)** because water and ethanol are (miscible) liquids with different boiling points. **(1)**

108. Carboxylic acids

1 (a) Its formula contains the –COOH/carboxyl group **(1)** and this is the functional group found in carboxylic acids. **(1)**

   (b) drawn formula for butanoic acid, e.g.

   (2 marks if all atoms and bonds are shown correctly, but 1 mark only if COOH is shown without one or more bonds)

   (c) butanol/butan-1-ol **(1)**

109. Core practical – Investigating combustion

Four from the following for 1 mark each:

- height of calorimeter above the flame
- size/height of flame
- lid or no lid on the calorimeter/use of insulation on sides of calorimeter
- starting temperature of the water/temperature of the room
- heating time
- draughts

110. Nanoparticles

1 Two advantages for one mark each, e.g. building stays warmer in winter/building stays cooler in summer/reduced energy costs / reduced use of fossil fuels / reduced carbon dioxide emissions.

111. Bulk materials

1 Copper is a good conductor of electricity. **(1)** Copper and PVC are ductile so they bend without breaking. **(1)** PVC is an insulator/poor conductor of electricity **(1)**, so you do not get electrocuted if you handle the wire **(1)**.

112. Extended response – Materials

Answer could include the following points **(6)**:

advantages of using aluminium alloy:

- stronger than ABS, so the phone will not bend so easily
- ductile, so easily shaped
- does not corrode easily (protected by an oxide layer)
- can be polished to an attractive shine
- denser than ABS, so may have a heavyweight luxury feel to it

disadvantages of using aluminium alloy:

- denser than ABS, so may be too heavy for some users
- more expensive/2.5 times more expensive than ABS
- Aluminium ores are a finite resource.
- a lot of electricity/energy needed to extract aluminium

advantages of using ABS:

- cheaper than aluminium alloy
- lower density than aluminium alloy so phone will be more lightweight
- can be moulded into shape
- can be coloured

disadvantages of using ABS:

- weaker than aluminium alloy, so body may need to be thicker to stop it bending
- Appearance may be less attractive than aluminium alloy.
- Addition polymers are made from crude oil, which is a finite resource.
- A lot of energy is needed to crack oil fractions to make alkenes for monomers.

other properties:

- Aluminium has a higher melting point than ABS.
- However, both temperatures are above the likely temperature at which the phone will be used.

conclusion (one material identified as better, with reasons given), e.g.

- Aluminium is better because it is stronger and more attractive, even though it is more expensive.
- ABS is better because it is cheaper and can be produced in different colours, which may appeal to different types of customer.

The Periodic Table of the Elements

1	2												3	4	5	6	7	0
																		4 **He** helium 2
7 **Li** lithium 3	9 **Be** beryllium 4												11 **B** boron 5	12 **C** carbon 6	14 **N** nitrogen 7	16 **O** oxygen 8	19 **F** fluorine 9	20 **Ne** neon 10
23 **Na** sodium 11	24 **Mg** magnesium 12												27 **Al** aluminium 13	28 **Si** silicon 14	31 **P** phosphorus 15	32 **S** sulfur 16	35.5 **Cl** chlorine 17	40 **Ar** argon 18
39 **K** potassium 19	40 **Ca** calcium 20	45 **Sc** scandium 21	48 **Ti** titanium 22	51 **V** vanadium 23	52 **Cr** chromium 24	55 **Mn** manganese 25	56 **Fe** iron 26	59 **Co** cobalt 27	59 **Ni** nickel 28	63.5 **Cu** copper 29	65 **Zn** zinc 30		70 **Ga** gallium 31	73 **Ge** germanium 32	75 **As** arsenic 33	79 **Se** selenium 34	80 **Br** bromine 35	84 **Kr** krypton 36
85 **Rb** ribidium 37	88 **Sr** strontium 38	89 **Y** yttrium 39	91 **Zr** zirconium 40	93 **Nb** niobium 41	96 **Mo** molybdenum 42	[98] **Tc** technetium 43	101 **Ru** ruthenium 44	103 **Rh** rhodium 45	106 **Pd** palladium 46	108 **Ag** silver 47	112 **Cd** cadmium 49		115 **In** indium 49	119 **Sn** tin 50	122 **Sb** antimony 51	128 **Te** tellurium 52	127 **I** iodine 53	131 **Xe** xenon 54
133 **Cs** caesium 55	137 **Ba** barium 56	139 **La*** lanthanum 57	178 **Hf** hafnium 72	181 **Ta** tantalum 73	184 **W** tungsten 74	186 **Re** rhenium 75	190 **Os** osmium 76	192 **Ir** iridium 77	195 **Pt** platinum 78	197 **Au** gold 79	201 **Hg** mercury 80		204 **Tl** thallium 81	207 **Pb** lead 82	209 **Bi** bismuth 83	[209] **Po** polonium 84	[210] **At** astatine 85	[222] **Rn** radon 86
[223] **Fr** francium 87	[226] **Ra** radium 88	[227] **Ac*** actinium 89	[261] **Rf** rutherfordium 104	[262] **Db** dubnium 105	[266] **Sg** seaborgium 106	[264] **Bh** bohrium 107	[277] **Hs** hassium 108	[268] **Mt** meitnerium 109	[271] **Ds** darmstadtium 110	[272] **Rg** roentgenium 111								

Elements with atomic numbers 112–116 have been reported but not fully authenticated

Key

relative atomic mass
atomic symbol
name
atomic (proton) number

1
H
hydrogen
1

*The lanthanoids (atomic numbers 58–71) and the actinoids (atomic numbers 90–103) have been omitted.

The relative atomic masses of copper and chlorine have been rounded to the nearest whole number.

Your own notes

Published by Pearson Education Limited, 80 Strand, London, WC2R 0RL.

www.pearsonschoolsandfecolleges.co.uk

Copies of official specifications for all Pearson qualifications may be found on the website:
qualifications.pearson.com

Text © Pearson Education Limited 2016
Typeset by Phoenix Photosetting
Illustrated by Techset Ltd.
Cover illustration by Miriam Sturdee

The right of Nigel Saunders to be identified as author of this work has been asserted by him in
accordance with the Copyright, Designs and Patents Act 1988.

First published 2016

21
10

British Library Cataloguing in Publication Data
A catalogue record for this book is available from the British Library

ISBN 978 1 292 13192 4

Copyright notice
All rights reserved. No part of this publication may be reproduced in any form or by any means (including photocopying or storing it in any medium by
electronic means and whether or not transiently or incidentally to some other use of this publication) without the written permission of the copyright
owner, except in accordance with the provisions of the Copyright, Designs and Patents Act 1988 or under the terms of a licence issued by the Copyright
Licensing Agency, Barnard's Inn, 86 Fetter Lane, London EC4A 1EN (www.cla.co.uk). Applications for the copyright owner's written permission should
be addressed to the publisher.

Printed in Great Britain by Bell and Bain Ltd, Glasgow

Acknowledgements
The author and publisher would like to thank the following for permission to reproduce photographs:

Science Photo Library Ltd: Gustoimages 95

All other images © Pearson Education

Notes from the publisher

1. In order to ensure that this resource offers high-quality support for the associated Pearson qualification, it has been through a review process by the
awarding body. This process confirms that this resource fully covers the teaching and learning content of the specification or part of a specification at
which it is aimed. It also confirms that it demonstrates an appropriate balance between the development of subject skills, knowledge and understanding, in
addition to preparation for assessment.

Endorsement does not cover any guidance on assessment activities or processes (e.g. practice questions or advice on how to answer assessment questions),
included in the resource nor does it prescribe any particular approach to the teaching or delivery of a related course.

While the publishers have made every attempt to ensure that advice on the qualification and its assessment is accurate, the official specification and
associated assessment guidance materials are the only authoritative source of information and should always be referred to for definitive guidance.

Pearson examiners have not contributed to any sections in this resource relevant to examination papers for which they have responsibility.

Examiners will not use endorsed resources as a source of material for any assessment set by Pearson.

Endorsement of a resource does not mean that the resource is required to achieve this Pearson qualification, nor does it mean that it is the only suitable
material available to support the qualification, and any resource lists produced by the awarding body shall include this and other appropriate resources.

2. Pearson has robust editorial processes, including answer and fact checks, to ensure the accuracy of the content in this publication, and every effort
is made to ensure this publication is free of errors. We are, however, only human, and occasionally errors do occur. Pearson is not liable for any
misunderstandings that arise as a result of errors in this publication, but it is our priority to ensure that the content is accurate. If you spot an error, please
do contact us at resourcescorrections@pearson.com so we can make sure it is corrected.